Demand Driven Supply Chain

Paulo Mendes Jr.

Demand Driven Supply Chain

A Structured and Practical Roadmap to Increase Profitability

Springer

Paulo Mendes Jr.
Catholic University
Industrial Engineering Department
Brazil
Author's contact: paulo.ddsc@gmail.com

ISBN 978-3-642-19991-2 e-ISBN 978-3-642-19992-9
DOI 10.1007/978-3-642-19992-9
Springer Heidelberg Dordrecht London New York

Library of Congress Control Number: 2011933105

Cover design: eStudio Calamar S.L.

Printed on acid-free paper

Springer is part of Springer Science+Business Media (www.springer.com)

To Betania, Matheus and Valentina, the Center of my Life

Acknowledgment

To my Lord, Jesus Christ who gave me the strength to continue in the most difficult periods of this journey and always sustain and guide my life.
To my parents, Paulo and Gloria, who strived to provide me the first education and always motivated me to continue my development. I am so thank you for your efforts.
To my friend and colleague, José Eugenio, who always provided good insight and great motivation to conclude this journey.
To my friends and professors Eugenio Epprech, John Vande Vate, Kleber Figueiredo, Peter Wanke and Andre Lacombe for all support and feedback provided to enrich the content of this book.
To my friend Lou Swanson, who provided me great motivation to finish this work.
To the Industrial Engineering Department of Catholic University (PUC) for all the knowledge provided during my MSc and PhD.

"If you Can Dream, you Can Do it"

Book Summary

Several companies have been implementing forecasting tools and processes to improve demand planning performance, but these initiatives were not enough to eliminate OOS problems, and improve supply chain efficiency, due to a mismatch between supply and demand, low forecast accuracy for medium and low volume products, high demand variability and/or high number of new product introductions.

To cope with this scenario, most companies are trying to move from a pure Push strategy, which is to produce and distribute based only on forecast, to a Pull system, which is to operate based on actual customer demand, in order to balance supply availability with customer demand.

This book aims to identify and describe the key components of demand driven supply chains, and based on these components, develop a structured and integrated assessment framework that companies can use to assess their current and desired future supply chain states in light of the Demand Driven Supply Chain (DDSC) concepts, and to define a supply chain strategy to move towards a customer centric operation, cost effectively.

The framework presented in this book was applied in different supply chain operations of a global CPG company to validate the methodology and formalize an action plan for these operations to be able to move towards a DDSC. The results of the assessment showed that two operations are currently close to a basic push level, and one is closed to an optimized push level, confirming that there are clear opportunities for those companies to improve their performance based on demand driven concepts.

Finally, another contribution of this book is the structured framework developed to design a 3-year supply chain strategy, which will consider the DDSC assessment results as one of the key inputs, and will support the implementation of the opportunities identified during the assessment.

Contents

About the Author

Paulo Mendes de Oliveira Junior Graduated in Nautical Sciences and Machinery from the Merchant Marine School in 1992. Has a PhD in Industrial Engineering from Catholic University, a Master of Science degree in International Logistics from the Georgia Institute of Technology, and a Master of Science in Industrial Engineering from Catholic University (PUC). Postgraduate in Marketing and in Management from Instituto de Administração e Gerência (IAG) of Catholic University (PUC), and in Finance from Escola de Pós-Graduação em Administração Financeira (EPGE) of FGV/RJ. Has global operational experience in the supply chain areas of Planning (Demand, Inventory and Operations and Production Planning), Procurement, Logistics and Commercial.

Chapter 1
Introduction

This chapter describes the goals, main contributions and areas covered by the book, and also reviews the current market and business environment which demands for a demand driven focus in order to reduce product out-of-stock levels and increase supply chain efficiency.

1.1 Book Overview

Companies in today's marketplace face an ever increasing requirement to improve customer service level, usually measured as order or case fill rate, to reduce product Out-of-Stock (OOS) on the shelf, at the same time reduce supply chain costs, defined as the sum of manufacturing, transportation, inventory, and distribution costs, to remain competitive in the marketplace.

However, when it is analyzed what actually happens in practice, it clearly be seen that most of the companies still experience high Out-of-Stock levels of critical products, as reported by several different studies, described below:

- Research done by Kraft foods and AC Nielsen in Brazil market in 2004 with 528 SKUs, 402 stores in São Paulo and 185 stores in Rio de Janeiro reported 8.0% OOS.
- Research done by a Beverage company in 2005 together with one of the biggest retail customer in the Brazil market reported an average OOS of 10.9%.
- Research done by Gruen (2007), reported a worldwide OOS level greater than 8%, as illustrated in Fig. 1.1.
- Research done by The University of Colorado, together with Grocery Manufacturers Association (GMA), Food Marketing Institute (FMI) and the National Association of Chain Drug Stores (NACDS), reported several root causes of Out-of-Stocks (OOS), being low demand forecast accuracy one of the critical causes.

Several companies have been implementing forecasting tools and processes to improve demand planning performance, but these initiatives have not been enough

P. Mendes, *Demand Driven Supply Chain*,
DOI 10.1007/978-3-642-19992-9_1, © Springer-Verlag Berlin Heidelberg 2011

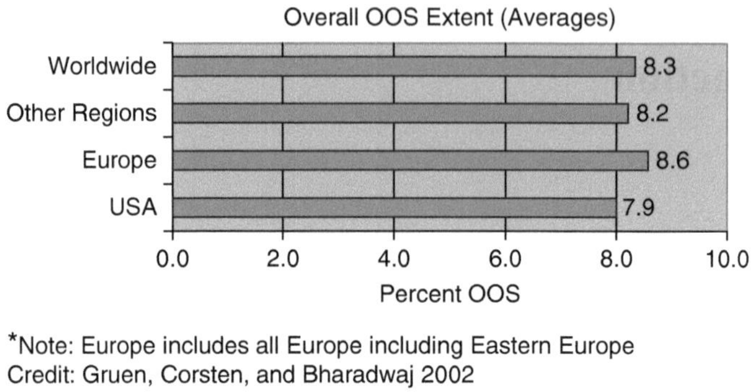

Fig. 1.1 Percent of OOS by geographical region (Gruen 2007)

to eliminate OOS problems, and improve supply chain efficiency, due to a mismatch between supply and demand, low forecast accuracy for medium and low volume products, high demand variability and/or a high number of new product introductions, which usually are much more difficult to predict than regular products. Kahn (2002) reported a study made with 53 products, from 16 firms in the US market that found a mean forecast error of 53% for new product forecasts. More recently, Jain (2007) reported a benchmarking study on new product forecasting, and as expected, the forecast error was 44% for products new to the company and to the world, and only 31% for products resulting from improvement of existing ones.

With low forecast accuracy and/or high demand variability, companies usually have to increase safety stock levels or transship products from one warehouse to another, on an expedite basis, when a warehouse is short of inventory, otherwise they will lose profit margin and become less competitive. However, these operational initiatives despite allowing companies to achieve the required service level, hurt operational efficiency and increase supply chain costs.

To cope with this scenario, many companies are trying to move from a pure Push strategy, produce and distribute based only on forecast, to a Pull system, operate based on actual customer demand, in order to better balance supply availability with customer demand, delivering the expected customer service level while, at the same time, achieving the required supply chain efficiency.

This book aims to provide three main contributions:

- A detailed and robust description of the concepts and components that makeup a Demand Driven Supply Chain.
- A structured and integrated framework companies can use to assess their supply chain in light of Demand Driven Supply Chain concepts.
- A supply chain strategy process to move towards a customer centric operation.

To accomplish these objectives, a detailed review of academic literature was performed to identify the components and characteristics of a demand driven supply chain, as these characteristics are not currently available in one single source.

The book is divided in eight chapters, starting in Chap. 2 with a review of the current academic literature available on Demand Driven Supply Chain concepts and assessment methodologies, and in Chap. 3, it is described the proposed framework to assess supply chains. In Chap. 4, it is reviewed and detailed the three key components of a Demand Driven Supply Chain. Based on the review made in Chap. 4, it is described the Demand Driven Supply Chain Model in Chap. 5, which consists of a five level maturity model that describes the characteristics of each supply chain functional process from a basic push operation level (level 1) to an optimized demand driven supply chain level (level 5), and in Chap. 6, the proposed Analytic Hierarchy Process (AHP) approach is presented to define weights for each component/category to be used in the assessment process. In Chap. 7, the assessment model is applied to three operations of a CPG Company to assess their current state, and also to validate the proposed model. In Chap. 8, results are summarized and also main conclusions identified after applying the methodology, as well as presented suggested areas for future developments.

1.2 Current Market and Business Environment

An article from The Economist Intelligence Unit (2009) reports that, due to economic uncertainty, volatile energy prices and intensifying global competition, large multinational corporations are seeking strategic and operational advantages, more than ever before. Among the key operational components, most demonstrably tied to business success, is the efficiency of global supply chains: the network of people, technology, activities, information and resources involved in supplying products or services to customers.

Velocity-based competition, less consumer loyalty, shortened product lifecycles, increased demand variability, globalization and global sourcing, leaner supply chains, more mass customization, and competitive pressures have altered the supply chain management requirements in fundamental ways, forcing organizations to rethink how they operate or risk being left behind.

The article also emphasizes that for years, the world's leading companies have been wringing inefficiencies out of their global supply chains, shrinking excess inventories and speeding order fulfillment, thus utilizing cash more efficiently, and better matching supply with customer needs. The principles of lean manufacturing and just-in-time inventory management famously helped Toyota leapfrog General Motors, as the world's largest vehicle manufacturer, and vaulted Wal-Mart to the forefront of global retailers, with US $405 billion in net sales revenue in 2010.

However, in today's fiercely competitive environment, it is not enough simply to streamline global supply chains and eliminate excess costs. Leading companies are applying new technologies and sophisticated analytics to make their supply chains more responsive to customer demand, rather than letting availability of supply drive the chain.

While planning remains a core part of any business, the faster pace of change not only in demand, but also supply and product, multiplies the problems that planning cannot always prevent. The financial impact to a company unable to respond to change can be crippling. Poor response can affect both the top line (e.g., inability to win new business, loss of customers to competitors, etc.), and bottom line (e.g., negative impact on margins, write-offs of excess and obsolete inventories, etc.).

At the end of 2007, and for the third year in a row in 2009, the 25 companies identified by Boston-based AMR Research (2009) as maintaining the top supply chains among the Fortune 500 enjoyed market-beating stock performance, with an average total return of 17.9% compared with 6.4% for the Dow Jones Industrial Average.

Over the past years, to effectively manage the volatility in demand, companies across a wide range of industries (e.g., automobile, fashion, etc.) have adopted demand-driven supply networks, using the "pull" of actual customer demand, rather than the "push" of available supply, to manage their network of suppliers, materials and components from manufacturing to distribution to improve supply chain efficiency while simultaneously meeting customer service requirements.

References

AMR Research Inc. (2009) The AMR Research supply chain top 25 for 2009. AMR Research, Boston

Gruen T (2007) Retail out-of-stocks: a worldwide examination of extent, causes, and consumer responses. University of Colorado, Colorado. http://www.uccs.edu/Igruen

Jain C (2007) Benchmarking new product forecasting. J Bus Forecast 28–29

Kahn K (2002) An exploratory investigation of new product forecasting practices. J Prod Innov Manage 19:133–143

The Economist Intelligence Unit (2009) The demand-driven supply chain: a holistic approach. The Economist Intelligence Unit, London

Chapter 2
Literature Review on Demand Driven Supply Chain (DDSC)

This chapter reviews the concepts of Demand Driven Supply Chain and the methodologies for assessing companies in light of DDSC concepts currently available in the academic literature, and finally, provides a summary of the operational and financial benefits of moving towards a DDSC.

2.1 Demand Driven Supply Chain Concepts

Before defining the DDSC concept, it is very important to review the concept of Supply Chain Management, as it will serve as the foundation to build the DDSC concept. To that end, the Council of Supply Chain Management Professionals (CSCMP 2009) defines Supply Chain Management as follow:

> ...Supply Chain Management encompasses the planning and management of all activities involved in sourcing and procurement, conversion, and all logistics management activities. Importantly, it also includes coordination and collaboration with channel partners, which can be suppliers, intermediaries, third party service providers, and customers. In essence, supply chain management integrates **supply** and **demand** management within and across companies...

And the boundaries and relationships are also defined as:

> ...Supply chain management is an integrating function with primary responsibility for linking major business functions and business processes, within and across companies, into a cohesive and high-performing business model. It includes all of the logistics management activities noted above, as well as manufacturing operations, and it drives coordination of processes and activities, with and across marketing, sales, product design, finance, and information technology...

Based on this definition, it can be pointed out two key concepts responsible for the success of Supply Chain Management initiatives in manufacturing and service companies: Supply Management and Demand Management.

Bayraktar et al. (2009) also confirm the importance of demand management. They tested a framework identifying the causal links among supply chain management

P. Mendes, *Demand Driven Supply Chain*,
DOI 10.1007/978-3-642-19992-9_2, © Springer-Verlag Berlin Heidelberg 2011

and information systems practices in small and medium size companies in Turkey. They performed hypotheses tests that indicate that both supply chain management and information systems practices positively and significantly influence the operational performance of 203 manufacturing companies considered in the analysis. One of the SCM practices identified was "close partnership with customers" or deemed by demand chain management.

Emmett and Crocker (2006) stated that Supply Chain Management is strategic and also operational. By strategic, they give an example that a company located in any one country needs to be thinking about global sourcing of raw material and packaging, new markets across the world, as the success of the business will ultimately depend on the success of this end-to-end supply chain of which the company is only part.

On the other hand, the supply chain is also operational, because the end-to-end supply chain concept has to work in practice, and this is all about getting supply chain thinking and skill-sets into every level of management and supervision, and into execution in every business function, in every player in the value chain. The drive for change needs to come from the top senior management, and the leadership of change to convert supply chain thinking into operational practice, must be taken up as a boardroom responsibility.

Emmett and Crocker (2006) stated that Logistics and Supply Chain are new concepts, emerging only in the 1980s and 1990s. They argue that supply has a connotation of being a push system, and for many the word "demand chain" is more meaningful, and that these concepts are being combined as "the Demand-Driven Supply Chain" (DDSC). They also explain that chains are being replaced by networks in an attempt to find new expressions to demonstrate how the thinking and practice can move forward.

Hull (2005) states that in a demand driven chain, a customer activates flow by ordering from the retailer, who reorders from the wholesaler, who reorder from the manufacturer, who reorder raw materials from the suppliers. Orders flow backward, up the chain, in this structure. The activator can be either actual customer demand as shown in Fig. 2.1, or forecasted customer demand.

AMR (AMR Research report 2005) defines the term "Demand Driven Supply Network" (DDSN) as a system of technologies and business processes that sense and

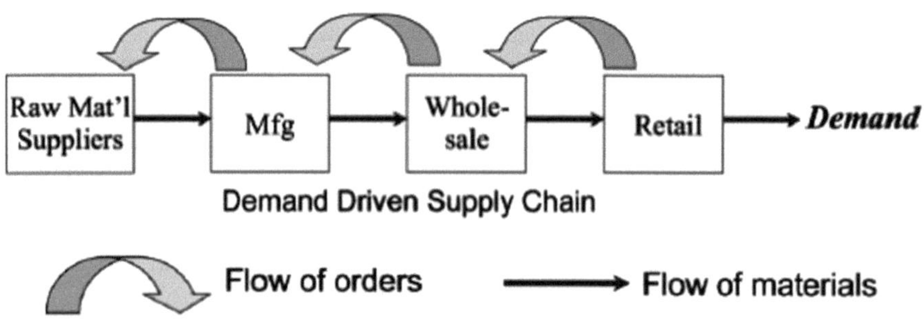

Fig. 2.1 Demand driven flow (Hull 2005)

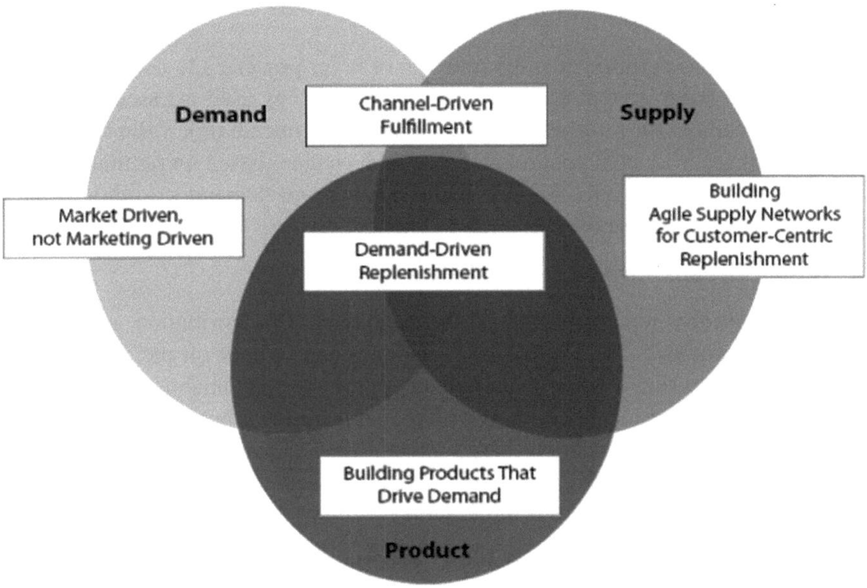

Fig. 2.2 AMR DDSN framework (AMR 2005)

respond to real-time demand across a network of customers, suppliers, and employees. The report also states that DDSN leaders are more demand sensing, which means being able to understand market drivers that impact demand, have more efforts for demand shaping, which means being able to influence the demand through specific market activities like special promotions, and focus on a profitable demand response.

AMR proposes five cross-functional strategies to become DDSN. These strategies are outlined on the AMR DDSN framework in Fig. 2.2:

- **Being Market driven and not Marketing driven:**

  In Demand Driven Supply Chain companies, processes are built from the outside-in, which means, they are based on a clear view of the customer, what is important for them and the requirements for account profitability. These companies become zealots on new product introductions and use their supply networks to shape and respond to demand.

- **Develop products that generate demand:**

  AMR argues that one of the successful factors of the AMR TOP 25 Supply Chain companies is excellence in innovation. Being quick to market with profitable products that are in high demand is a core competence of a DDSN strategy. For DDSN leaders, innovation excellence is a key to success, and it is infused into all supply chain processes. AMR research shows that 75% of new products fail, and **42% of companies lack a common set of internal standards for managing New Product Development and Introduction process**.

- **Have a Channel-driven fulfillment process:**

  Channel-driven fulfillment is the redesign of order processes to become demand driven, not order driven, and the supply chain strategy used is based on service level agreement for pull-based replenishment to define an order. Replenishment decisions are evaluated continuously for each channel based on profitability and product placement goals. Supply chain velocity and demand visibility are key elements for a successful execution of channel driven.

- **Have a Demand-driven replenishment process**

  Demand-driven replenishment is the alignment of distribution and manufacturing processes for a pull-based response, and is built on the principles of lean manufacturing – waste reduction and pull-based replenishment. It connects these principles of local execution with global planning process using pull-based network design and constraint-based planning in Sales and Operations Planning (S&OP). These principles are closely linked to manufacturing, procurement and logistics decisions in building agile networks.

- **Have Agile networks for a customer-centric response**

  Agile networks are built to align materials suppliers, contract manufacturers, and logistics providers to a demand signal. An agile network starts with the design of the network for pull-based replenishment, and is continually refined through New Product Development and Introduction processes.

  Agile networks start with the design and flexibility based on joint agreements (contract relationships and demand visibility are essential). The key elements of agility and reliability are balanced with cost for the selection of manufacturing sites, supplier qualifications and modes of transportation.

  Ayers and Malmberg (2002) describe a Demand-Driven Supply Chain as one in which the company is trying to shift from "build to forecast" to "build to order" discipline. The Demand-Driven Supply Chain is one that derives the information for production and inventory decisions from actual, real-time demand, and not forecasts – even if the forecasts use past sales history as a basis. They also argue that the property of being demand-driven is one of degree:

- Being "zero percent" demand-driven, means all production/inventory decisions are based on forecasts, and so, all products available for sale to the end user is there by virtue of a forecast. This could be the case of fashion goods, where the designer may not know how buyers will react to a new design, or the beverage industry, where products are produced based on a given forecast.
- A "100 percent" demand-driven is one in which the order is received before production begins. In this model, the commercial aircraft industry comes close to this description.

Bowersox and Lahowchich (2008) describe that traditional supply chains have been designed to operate in an anticipatory, or a "push mode." The prevailing distribution process is a time-consuming, forecast driven, volume oriented, functionally centric consolidation process designed to "push" products to market destinations in anticipation of future demand. The frequent result of this anticipatory push process is far too much of the wrong inventory being pushed to the wrong markets, and this missed alignment of inventory often results in firms using incentives to entice consumers to buy products they have available to sell, rather than providing the exact product the consumers desire to purchase.

Throughout different industrial segments, business leaders and consultants had difficulty explaining why, at the end of the week, or month, despite inventories reaching high levels, out-of stocks were excessive. It is also difficult to fully understand why 70–80% of trade sales of some consumer products like beverage, food, disposable diapers, occur in the last week of the month or at the end of the business quarter. The chart in Fig. 2.3 gives an example of the high sales variability due to promotions faced by a beverage company in one region of Brazil.

The reason for such sales concentration is that companies are required to meet monthly, quarterly or annually sales goals, and to achieve these objectives, they provide incentives in the form of product promotions or price discounts in order to achieve a lift in the customer demand and therefore, meet the required business objectives. This focus on "sell in" to the customer, instead of focus on "selling out" from the customer to the consumer, increases inventory levels, but does not reduce out-of-stock implications, as more frequently, the products that receive incentives are those with high volume impact, and out-of-stock usually happens in products with low volume impact. This type of characteristic

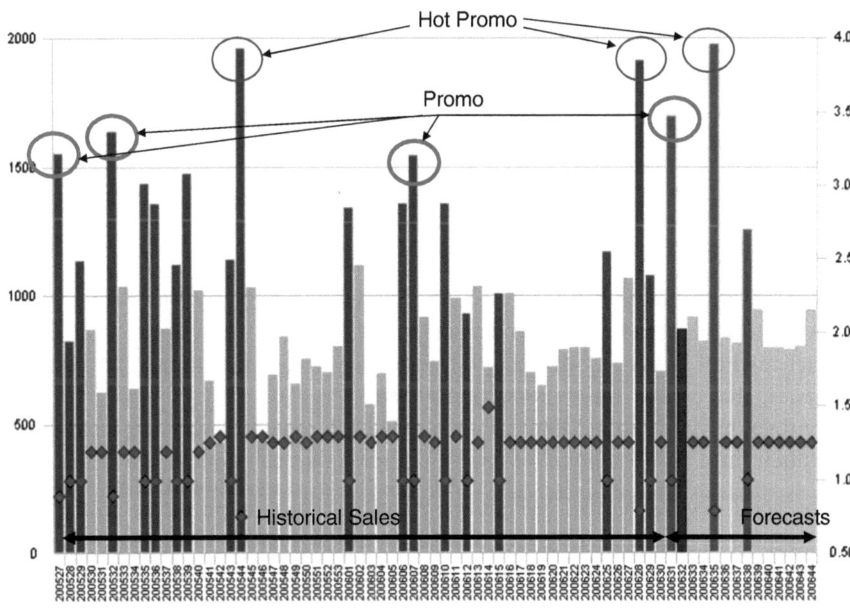

Fig. 2.3 Weekly sales volume of a Brazilian Beverage Company

is frequently found in sales driven companies, which should not be confused with demand driven companies.

Michael L. Eskew, recently retired chairman and CEO of UPS, presented the overall transformation challenge from the perspective of global companies and their service providers at the 2007 Longitudes conference:

> ...Consumer pull requires one-to-one solutions and supply chains that can deliver them. The world is no longer driven by producers pushing products through their supply chain. Increasingly, power is in the hands of consumers who now pull products through the system. They pull what they want, when they want it, from whomever they choose anywhere in the world, and consumers want and expect a personal, relevant, individualized experience, and this is a big shift that will only intensify...

As there are many definitions of Push vs. Pull systems, it is important to clear define the two concepts, as they are keystone in the demand driven concept.

At the 2005 Aspen Institute Roundtable on Information Technology, 25 thought leaders discussed the broad implications of push vs. pull economies, generating the following definitions:

- A push economy is the kind of economy that was responsible for mass production in the twentieth century, and is based on anticipating consumer demand, and then, making sure that needed resources are brought together at the right place, at the right time, for the right people. A company forecasts demand, specifies in advance the necessary inputs, regiments production procedures, and then pushes the final product into the marketplace using standardized distribution channels and marketing.
- A pull economy is the kind of economy that appears to be materializing in online environments, and is based on open, flexible production platforms that use networking technologies to orchestrate a broad range of resources. Instead of producing standardized products for mass markets, companies use pull techniques to assemble products in customized ways to serve local or specialized needs, usually in a rapid or more informal, "on-the-fly processes."

Hopp and Spearman (2003) provide a brief history of the Pull system and also a more clear definition of strategic and tactical Pull system, as well as Push system:

- Strategic Pull can be defined as establishing a takt time to set the output of the production plant to be equal to demand.
- Tactical Pull system is the one that explicitly limits the amount of work in process that can be in the production system.
- By default, it is implied that a Push production system is the one that has no explicit limit on the amount of work in process that can be in the system.

The good news about this definition of Pull is that it implies that pull can be implemented in a variety of ways. To illustrate this argument, Hopp and Spearman (2003) give some examples of common systems found in industry and how they should be classified in either Push or Pull, as detailed below:

- MRP is a push system because releases are made according to a master production schedule without regard to system status. Hence, no a priori work in process (WIP) limit exists.
- MRP with a WIP constraint is a pull system.
- Classic Kanban is a pull system, as the number of kanban cards establishes a fixed limit on WIP.
- Classic Base Stock System is a push system because there is no limit on the amount of work in process in the system.
- Installation stock (Q,r) is a push system as it does not impose a limit on the number of orders in the system.

They also argue that there are three primary logistical reasons for the improved performance of pull systems:

- Less congestion – Comparison of an open queuing network with an "equivalent" closed one shows that the average WIP is lower in the closed network than the open network given the same throughput.
- Easier control – WIP is easier to control than throughput since it can be observed directly.
- WIP Cap – The benefits of a pull environment owe more to the fact that WIP is bounded than to the practice of "pulling" everywhere.

Ashayeri and Kampstra (2005) also provide a concise definition, as described below:

- PUSH – Node performs order planning for succeeding node. Control information flow is in the same direction of goods flow.
- SEMI PUSH or PUSH–PULL – Succeeding node makes order request for preceding node. Preceding node reacts by replenishing from stock that is rebuilt every fixed period.
- PULL – Succeeding node makes order request for preceding node. Preceding node reacts by producing the order, which involves all internal operations, and replenishes when finished.
- SEMI PULL or PULL–PUSH – Succeeding node makes order request for preceding node. Preceding node reacts by replenishing from stock that is rebuilt immediately.

Harrison (2003) describes three different supply chain strategies that a company can implement:

- Push-based strategy in which production and distribution decisions are based on long-term forecasted demand. In this case, it takes much longer to the company to react to the changing marketplace. As the strategy relies on forecasts, it is most of the time difficult to match supply and demand.
- Pull-based strategy in which production and distribution are demand driven, so that they are coordinated with true customer demand rather than forecast. In this case, the company does not hold any inventory and only produces to order. These systems are intuitively attractive since they allow the company to

eliminate inventory while responding to customer demand. Unfortunately, it is very difficult to implement a pull based strategy when lead times are so long, that it is impractical to react to demand information. Similarly, it is frequently more difficult to take advantage of economies of scale, since production and distribution decisions are made in response to specific customer demand, and therefore, batch production or efficient transportation modes, such as truckloads, are hard to achieve.

The advantages and disadvantages of Push and Pull supply chain strategies have led companies to look for a new supply chain strategy that exploits the best of both worlds: The Hybrid Push–Pull supply chain strategy.

- Hybrid Push–Pull strategy in which some stages of the supply chain, typically the initial stages, are operated in a Push-based manner, while the remaining stages are operated in a Pull-based strategy, and the interface between the Push-based stages and the Pull-based stages is usually referred to as the "Push–Pull boundary."

Harrison (2003) also argues that the challenge for the firms is to define which of the three supply chain strategies described above is most appropriate for each product. Figure 2.4 below provides a framework to match supply chain strategies with products and industries. In the vertical axis, it is shown information on uncertainty in customer demand, while the horizontal axis represents the importance of economies of scale, either in production or distribution:

Assuming everything else being equal, the higher the demand uncertainty, the more the firm would prefer managing the supply chain based on realized demand, that is, based on a Pull strategy. On the other hand, the smaller the demand uncertainty, the more the firm would be interested in managing the supply chain based on forecast, that is, based on a Push strategy. The same logical is true for analyzing the economies of scale, that is, the higher the importance of economies of scale

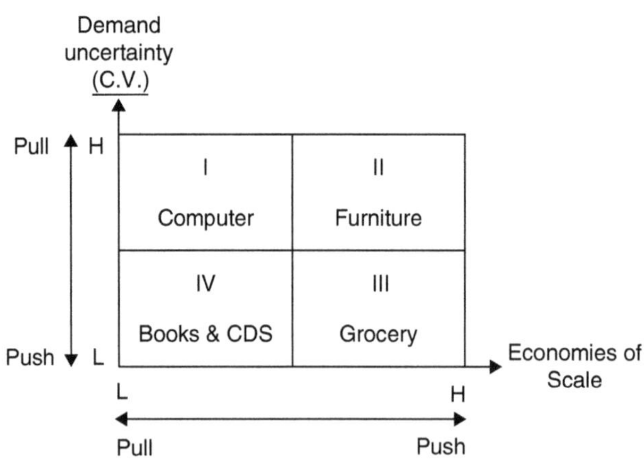

Fig. 2.4 Matching supply chain strategies with products (Harrison 2003)

in reducing cost, the more important is to aggregate demand, and thus, the more important is to manage the supply chain based on forecast.

Based on the framework illustrated in Fig. 2.4, Harrison (2003) summarizes when to use each one of the three supply chain strategies:

- Push based supply chain strategy, usually suggested for products with small demand uncertainty, as the forecast will provide a good direction on what to produce and keep in inventory, and also for products with high importance of economies of scale in reducing costs.
- Pull based supply chain strategy, usually suggested for products with high demand uncertainty and with low importance of economies of scales, which means, aggregation does not reduce cost, and hence, the firm would be willing to manage the supply chain based on realized demand.
- Hybrid Push–Pull strategy, usually suggested for products which uncertainty in demand is high, while economies of scale are important in reducing production and/or delivery costs. One good example of this strategy is the furniture industry, where production strategy has to follow a Pull-based strategy, since it is impossible to make production decisions based on long-term forecasts. On the other hand, the distribution strategy needs to take advantage of economies of scale in order to reduce transportation cost, using a Push-based strategy.

For a hybrid Push–Pull strategy, a second important decision is to define where to locate the Push–Pull boundary in the supply chain. Harrison (2003) states that the Push part is applied to the portion of the supply chain where demand uncertainty is relatively small, and thus, managing this portion based on long-term forecast is appropriate. On the other hand, the Pull part is applied to the portion of the supply chain where uncertainty is high, and hence, it is important to manage this part based on realized demand. One illustrative example is Dell, who implemented the Push–Pull strategy by locating the boundary at the assembly point.

Wanke et al. (2010) argue that the perception of logistics systems being complex is confirmed by several authors, but it is not always clear what does it mean. They defined complexity in logistics in terms of quantifiable scales and based on the notion of numerous actors or parts that are interconnected and can be captured by measures such as the company's gross revenue, its number of suppliers, active customers, number of employees, number of employees involved in supply chain management, active stock keeping units (SKUs), number of distribution centers, orders processed and new product launches per year.

They proposed that logistics complexity is a driver to define the way a company manages and emphasizes the different supply chain objectives and decision areas, and based on this, a contingency approach for supply chain management is required, where different contextual conditions drive the way the supply chain choices are made and management activities are performed, as opposed to a best practice approach where there would be some universally applicable principles that would be appropriate regardless of the particular conditions under study.

Zeithaml et al. (1988) describe that the essential premise of the contingency approach is that effectiveness, broadly defined as organizational adaptation and survival can be achieved in more than one way. They give the example that there is more than one way to organize effectively, and more than one leadership style that can achieve organizational goals. The contingency approach therefore, suggests that it can be observed wide variations in effectiveness, but that these variations are not random. Effectiveness depends on the appropriate matching of contingency factors with internal organizational designs that can allow appropriate responses to the environment.

One example of contingency approach applied to supply chain management comes from Fisher (1997). He proposes a framework to define what is the best supply chain for a company's product. He argues that the first step in devising an effective supply chain strategy is to consider the nature of the demand for the products. To that end, many aspects are important, for example, product lifecycle, demand predictability, product variety, and market standards for lead time and service. He proposes to classify products on two categories: They are either primarily **Functional** or primarily **Innovative**, as summarized below:

Functional Products:

- Product do not change much over time;
- Have stable and predictable demand;
- Long life cycles;
- Lower potential growth.

Innovative Products:

- Great variety of products;
- Increase unpredictability (volatile demand);
- Short life cycles;
- Higher potential growth.

The next step should be to decide whether the company's supply chain is "**Physically Efficient**" or "**Responsive to the Market,**" as described in Table 2.1 below.

After determining the nature of the product demand and the supply chain priorities, managers can employ a matrix to formulate the ideal supply chain strategy. Fisher proposes to plot the nature of the demand for each of the product

Table 2.1 Physically efficient vs. market responsive supply chains (Fisher 1997)

	Physically efficient process	Market responsive process
Primary purpose	Supply predictable demand efficiently at the lowest possible cost	Respond quickly to unpredictable demand in order to minimize stock outs and obsolete inventory
Manufacturing focus	Maintain high average utilization rate (reduce setups)	Deploy excess buffer capacity
Inventor strategy	Generate high tums and minimize inventory throughout the chain	Deploy significant buffer stocks or end products in the chain
Lead time focus	Shorten lead time as long as it does not increase cost	Investaggressively in ways to reduced lead time
Approach to choosing suppliers	Select primarily for cost and quality criteria	Select primarily for speed, flexibility, and quality
Product-design strategy	Maximize performance and minimize cost	Try to postpone product differentiation for as long as possible in the supply chain

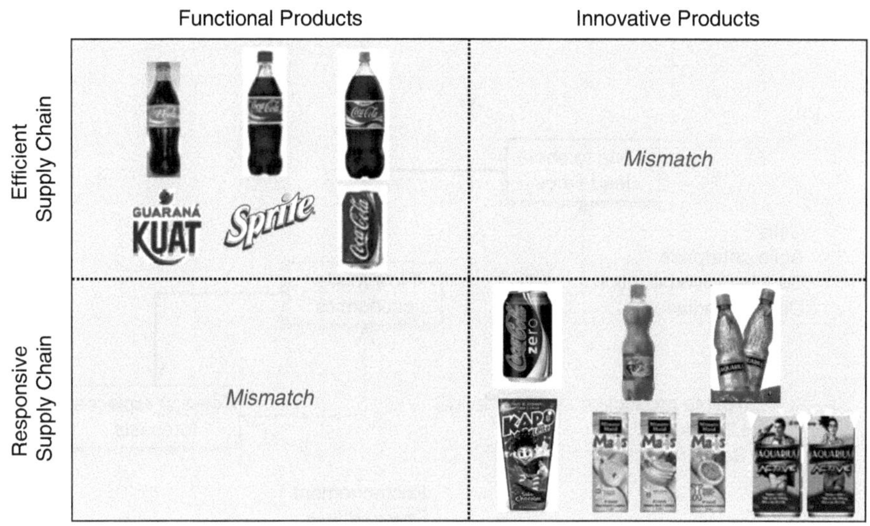

Fig. 2.5 Example of product and supply chains applied to beverage industry

families and its supply chain priorities, in order to allow identify whether the process used for supplying products is well matched to the product type, which means, an efficient process for functional products and a responsive process for innovative products. In Fig. 2.5, the author shows an example of the proposed matrix applied to a practical case in the beverage industry.

Ayers (2006) advocates that the Demand-Driven Supply Chain changes many of the conditions that cause wasteful variation in supply chain production. He states that, it is the foundation of the "lean" supply chain, and its implementation helps establish the operating range for low-cost production supply chain.

He proposes a three-phase roadmap to implement the Demand-Driven Supply Chain concept. The phases are listed and also illustrated in Fig. 2.6:

1. **Moving from long to short lead-times** – Overall lead-time is composed of individual cycle-times for multiple processes. This step involves shortening the cycle-time at each step in the critical path processes from the point of purchase to the start of production for the entire supply chain.
2. **Replacing the batch with the flow model economics** – Flow model economics encompass low-cost ways to vary mix and volume. Lean manufacturing is a discipline that has the same goals as flow economics. Examples include "**single minute exchange of dies**" (**SMED**) in manufacturing, which will be specified in Chap. 5, and mixing different products on production lines. Batch picking for multiple customers in a warehouse would represent a non-manufacturing example. A flow model will synchronize supply chain steps and increase the overall supply chain ability to respond to changes.
3. **Basing decisions on actual demand rather than forecasts** – This step requires efficient sharing of information up and down the chain. An ideal process is to

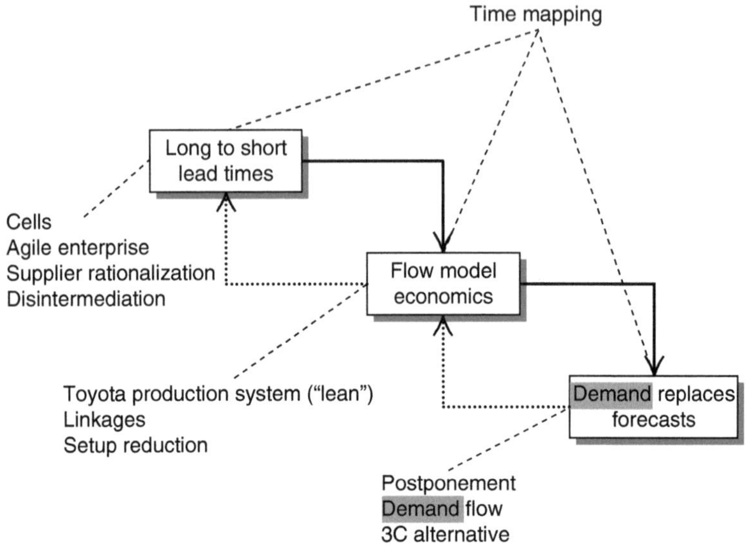

Fig. 2.6 Three phase roadmap for implementing a DDSC (Ayers 2006)

have all supply chain partners with access to real time sales, as well as, to the business rules to react based on demand signal.

Evolution to a demand-driven supply chain will likely proceed in the order proposed above. Shortening the lead-time is fundamental to changing batch model economics. Basing decisions on demand comes after adopting the economics of the flow model. Along the path, there is feedback to earlier steps. For each phase in Fig. 2.6 above there are three to four methodologies to be applied towards a DDSC operation.

Another key concept related to Demand-Driven Supply Chains is the concept of Agile. Christopher (2000) presents the concept of agility as a business wide capability that embraces organizational structures, information systems, logistics processes, being flexibility one of the key characteristics. He also identifies the four characteristics of an agile supply chain as market sensitive, network based, process integration and virtual supply chains, being this last characteristic defined as the information sharing network between buyers and suppliers.

He explains the difference between **agile** and **lean** concepts. He defined Lean as doing more with less, and explains that the term is often used in connection with lean manufacturing to imply a zero inventory approach. However, there are certain conditions where a lean approach to supply chain makes sense, in particular where demand is predictable and the requirement for variety is low and volume is high. The problem arises when attempting to implant the lean concept into situations where demand is less predictable, the requirement for variety is high and the volume at the individual SKU (Stock Keeping Unit) level is low, which is the regular characteristics of several markets and products around the world.

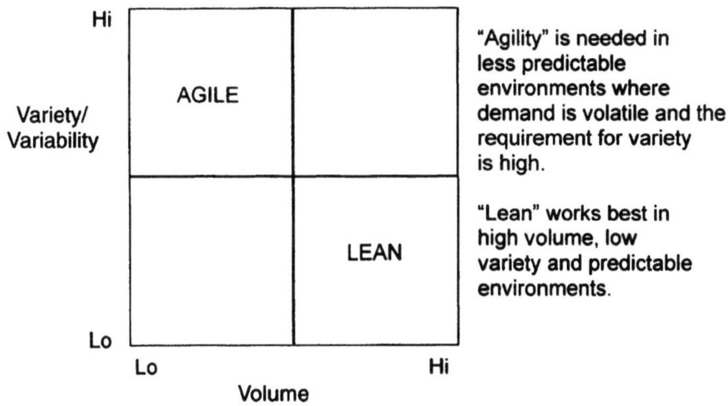

Fig. 2.7 Characteristics of Agile and Lean concepts (Christopher 2000)

On the other hand, agility is defined as the ability of an organization to respond rapidly to changes in demand, both in terms of volume and variety. Figure 2.7 summarizes the three critical dimensions – variety, variability (or predictability) and volume – that determine which approach – agile or lean – should be deployed.

Christopher (2000) also states that to be truly agile, a supply chain must possess four distinguishing characteristics, being one of them **Market sensitive,** which means that the supply chain is capable of reading and responding to real demand or being demand-driven. The problem is that most organizations are forecast-driven rather than demand-driven. In other words, because they have little direct feed-forward from the marketplace by way of data on actual customer requirements, they are forced to make forecasts based on past sales or shipments, and convert these forecasts into inventory.

One of the biggest barriers to agility is the way that complexity (product and brand proliferation, organizational structure and management processes) tends to increase as a company grows and expands its market coverage. The reduction of complexity should be a major priority for Marketing and Logistics functional areas to allow a company become agile.

Agarwal et al. (2007) review the literature on supply chain agility, touching some of the components of DDSC like lead time reduction, market sensitiveness, new product introduction, and propose to apply Interpretive Structural Modeling (ISM) to show the interrelationship of different 15 variables to supply chain agility.

Huang et al. (2009) propose an agile approach for supply chain modeling using a generic label correcting (GLC) algorithm. The rough set theory, which is a mathematical approach to manage vague and uncertain data or problems related to information systems, indiscernible relations and classification, is applied to reduce the complexity of data space when running the algorithm.

Ismail and Sharifi (2006) present a structured framework to provide a practical approach for implementing agile supply chains (ASC), based on the concepts of supply chain design and design for supply chain.

2.2 Methodologies for Assessing Demand Driven Supply Chain

In terms of methodologies to assess and identify company's performance, several articles show the importance of having a structured process in order to improve performance overtime. However, when it is specifically related to methods for assessing performance based on Demand Driven Supply Chain (DDSC) concepts, the articles available do not meet the research criteria which is to have a clear and practical framework to support companies identify their current state based on DDSC concepts.

Dale and Ritchie (2000) argue that companies must have an appropriate performance measurement system to be applied on a regular basis to identify areas to be improved in order to establish a sustainable continuous improvement process. They proposed to use self-assessment process, which can be defined as a comprehensive, systematic and regular review of an organization's activities and results against a model of business excellence. The self-assessment will allow organizations to clearly discern its strengths and gaps, and define improvement actions linked to the business planning process.

They state that there are some necessary criteria for a successful self-assessment process:

- Gaining commitment and support from all levels of staff
- Action being taken from the previous self-assessment
- Incorporation of self-assessment into the business planning process
- Not allowing the process to be "added on" to employees existing workload
- Developing a framework for performance monitoring

In terms of benefits of the self-assessment, there are both immediate and long term benefits:

Immediate benefits:

- Facilitates benchmarking, drives continuous improvement, encourages employees involvement and ownership, provides visibility in direction, raises understanding and awareness of quality related issues, develops a common approach to continuous improvement across the company.

Long term benefits:

- Keeps costs down, improves business results, provides a disciplined approach to business planning, increases the ability to meet and exceed customers' expectations.

Chin et al. (2003) also developed a knowledge-based expert self-assessment (KES) training toolkit to measure and assess organizational performance based on the evaluation criteria of the renowned business excellence model – The Malcolm Baldrige National Quality Award (MBNQA).

The concept of self-assessment brings a valuable contribution to reduce complexity, time and cost to apply the DDSC assessment framework on a global basis, as each company should be able to self-assess its current state.

Salama et al. (2009) review the importance of supply chain and operations audit process which represents a fundamental step to support improvement projects. They argue that the core element of audits is the diagnostic stage and that no audit can be considered successful unless it really provides a thorough understanding of how the constituent elements of an organization interact with one another (e.g., people, processes and technologies), that is the interactions which constrain the system, and how these interactions are reflected on the market-driven performance. The provided a very clear set of features and requirements for an audit methodology that can be considered when developing a DDSC assessment:

- Quick/Accurate – The methodology should be based on tools, steps and an "engine" which were designed to deliver a result as accurate as possible in the shortest time possible.
- Not invasive – The methodology should be built in order to require the least possible effort from organization's resource.
- Scalable – The methodology should be scalable.
- Avoid bias/theoretically grounded – The methodology should be built in a way to reduce possible bias in the diagnostic stage, while exploiting the knowledge that people who daily work in an organization have on their processes.
- Stimulate consensus building – The stimulation of consensus building can be achieved in different ways. The most important are:
  - Possible recycles in the diagnostic stage
  - Empirical support of critical findings
  - Quantification of value together with scenario analysis
- Transparent – All tools and steps used in the methodology should be clearly described in all parts. No "secret engine" is behind the methodology.

The proposed new audit methodology by Salama et al. (2009) were tested through three European research initiatives, and also showed an example of a master best practice relationship map for the demand management process.

Moon (2002) also provides direction on the importance of auditing process related to sales forecasting. He states that sales forecasting audit process has three objectives:

- Understand current status of forecasting practice (a company's "as is" state).
- Visualize the goals of forecasting process improvement (the "should-be" state).
- Develop a roadmap for achieving the goals (the "way forward").

Trkman and McCormack (2009) describe that supply risk or supply disruptions is emerging as a key challenge to supply chain management, and that the ability to identify which supplier has greater potential of disruption is a critical step in managing the frequency and impact of these disruptions.

Their contribution was to use the contingency theory approach to propose a new method for the assessment and classification of suppliers based on their supply chain characteristics, its structure and supplier's attributes and performances, modified by factors in the supplier's specific environment namely exogenous and endogenous uncertainty. The contingency approach is a value contribution to be considered when developing the DDSC assessment framework as different companies and industries can have different time and market requirements to move or not to move towards DDSC.

Filho et al. (2010) developed a framework to measure safety culture in the Brazilian oil and gas companies. They applied a five level safety culture maturity model (e.g., pathological, reactive, bureaucratic, proactive and sustainable) using five dimensions (e.g., information, organizational learning, involvement, communication and commitment) to identify current state of safety practices in petrochemical companies.

A maturity model can be described as a structured collection of elements that describe certain aspects of maturity in an organization, and aids in the definition and understanding of the different organization processes. A maturity model can be a valuable tool to describe the different maturity levels in the DDSC assessment process.

One of the key objectives of DDSC is to reduce demand amplification as it brings extra costs and inefficiencies like extra resource capacity, higher inventory levels, etc. Taylor (2000) reviews the effect of demand amplification in the supply chain and also proposes a practical approach to eliminate it through a seven step process. A pilot test was performed in UK automotive industry and showed an increase from 70 to 100% on the composite measure of delivery to time along the supply chain, and also a reduction of 30% in total supply chain inventory.

Childerhouse et al. (2002) proposed a methodological framework to develop focused demand chain strategy for each cluster of products commercialized by a company. The methodology consists of six steps described below and has the objective to define the best facility, production layout requirements and control mechanisms for each specific product/service offered by the company.

- Step 1: Develop holistic demand chain strategy. This leads from highlighting of core competencies and resources, and its primary purpose is the identification of specific markets to be targeted plus the overall corporate strategy.
- Step 2: Identify specific product/service offering. These are tailored to the target markets with emphasis placed on prioritization of service, quality, cost or lead times.
- Step 3: Categorize demand chain types. Given the specific products and their related service criteria, the DWV3 classification variables (duration of lifecycle, time window for delivery, volume, variety and variability) are used to categorize the products into clusters with similar characteristics. Output is a clear definition of the requirements for each demand channel.
- Step 4: Identify facility requirements. Facilities need to be tailored to achieve the desired objectives (e.g., products with high service level may require distribution warehouses located near the marketplace).
- Step 5: Define production layout and control mechanisms (e.g., Kanban, MRP, etc.).
- Step 6: Implement focused demand chains.

The proposed methodology was applied to a UK lighting company and showed several benefits like 75% reduction in product development time, 27% reduction in manufacturing costs, and 95% reduction in delivery lead times.

Bowersox and Lahowchich (2008) propose a Responsive Supply Chain Business Model and describe it as a "*…customer-facing organization and operational strategy focusing the highest priority on providing exacting and sustainable customer service…*". They explain that the Responsive Supply Chain business model represents a blend of six imperatives, or essential elements (1) Consumer connectivity, (2) Operational excellence, (3) Integrative management, (4) Real-time responsiveness, (5) Leveraging networks, and (6) Collaboration, and each of these six imperatives represents a firm's unique supply chain DNA.

Verdouw et al. (2010) analyzed the European fruit market and identified that fruit supply does not sufficiently meet demand requirements. They proposed that the fruit supply chains needed to become demand driven, that is, being able to continuously match supply capabilities to changing demand requirements. In a demand driven supply chain, all actors involved are sensitive and responsive to demand information of the ultimate consumer and meet those varied and variable demands in a timely and cost-effective manner. As a consequence, information must be shared timely throughout the supply chain and the early alerted firms have to respond quickly to changes in demand or supply, which imposes stringent demands on the interoperability and flexibility of the enabling information systems.

They presented a reference model for designing business processes in demand-driven fruit supply chains. The model consists of a reference modeling framework that defines process models at different levels of abstraction and includes a method of how they can be composed from a repository of building blocks. However, they did not provide any structured assessment approach to evaluate different business segments/industries in light of demand driven supply chain concepts.

Georgiadis et al. (2001) present a paper describing the design and implementation of a demand driven freight transport application, but they focus mainly in the IT system architecture of the solution, called ATLog (Attika Traffic Logistics), not providing any direction on how to assess and determine a company current state based on DDSC concepts.

Ayers and Malmberg (2002) touch very briefly DDSC concepts, providing a four stage maturity model to show how enablers of supply chain improvement support the introduction of information technology to the supply chain, and one of these elements is the demand-driven as illustrated in Fig. 2.8 below. However, they did not provide a detail maturity model and a robust methodology to assess a supply chain, in order to determine its current state in terms of the demand driven concepts.

Table 2.2 below provides a summary of the current literature review on assessing DDSC:

Exhibit 3. ***Enablers of Supply Chain Improvement***

	Stages of Supply Chain Evolution			
	1	2	3	4
	Infrastructure	**Cost Reduction**	**Collaboration**	**Strategic Contribution**
Supply Chain Organization	Capability building Execution of basic tasks	Root cause analysis Item stratification "Optimal" buys	Intercompany improvement programs Financial sharing arrangements	Customer-focused organization structures along supply chain Supply chain level steering committees
Demand-Driven Supply Chain	Supplier reduction Negotiations Mostly forecast driven; backorder and simple service level indicators	Quality systems Outsourcing Supplier ratings Lead-time reduction (cells, small batches) Modest JIT/pull arrangements; 25 % of demand-driven potential	Design changes Postponement strategies Information exchange (inventories, forecasts, demand) 50 % application of demand-driven potential	Segment strategies New product involvement 80–90% implementation of demand-driven potential

Fig. 2.8 Enablers of supply chain improvement (Ayers and Malmberg 2002)

Table 2.2 Summary current literature review on DDSC

Author(s)	Contribution
Dale and Ritchie (2000)	Proposed to use self-assessment process to evaluate company's performance on a regular basis as part of the continuous improvement process
Chin et al. (2003)	Developed a knowledge-based expert self-assessment (KES) training toolkit to measure and assess organizational performance
Salama et al. (2009)	Review the importance of supply chain and operations audit process
Moon (2002)	Provides direction on the importance of auditing process related to sales forecasting
Trkman and McCormack (2009)	Argue that supply risk is one of key challenge to supply chain management and propose a new method for assessment and classification of suppliers
Filho et al. (2010)	Developed a five level maturity model to measure safety culture in the Brazilian oil and gas companies
Taylor (2000)	Reviews the effect of demand amplification in the supply chain and also proposes a seven step process to eliminate it
Childerhouse et al. (2002)	Proposed methodological framework to develop focused demand chain strategy for each cluster of product commercialized by a company
Bowersox and Lahowchich (2008)	Propose a responsive supply chain business model
Verdouw et al. (2010)	Proposed a reference model for designing business processes in demand driven fruit supply chain in Europe
Georgiadis et al. (2001)	Describe the design and implementation of demand driven freight transport application
Ayers and Malmberg (2002)	Provide a four stage readiness model to show how enablers of supply chain improvement support the introduction of information technology to the supply chain

In this book, it is proposed to define the components of DDSC, then develop a structured methodology that will help companies assess their current state in light of demand driven supply chain concepts and identify their current strengths and gaps, and therefore, define a strategic plan to evolve and become more efficient and competitive.

2.3 Benefits of Demand Driven Supply Chain

Despite of the limited information available on the benefits of becoming demand driven, two different studies provide a direction on the financial and operational benefits companies can capture when implementing a demand driven supply chain.

Based on internal benchmark data, AMR reports that the most advanced demand-sensing companies have 15% less inventory, 17% better perfect order performance, and 35% shorter cash-to-cash cycle time. In terms of top line results, DDSC leaders have 10% higher revenue and 5–7% better profit margins than their competitors. These extraordinary results captured by demand-driven companies, show the importance of having a structured methodology for assessing the current state against DDSC concepts, in order to help companies evolve in the implementation of DDSC components and tactics.

Another reference comes from SAP (SAP Insight report 2006), which argues that based on existing customer studies, analyst comments and industry pooling, the implementation of DDSC can generate the following results:

Revenue: Increase fill rates (defined as cases delivered divided by cases ordered) and reduce out-of-stocks by 3–10%:

Operating cost:

- Increase production efficiencies by 1–5%
- Decrease freight costs by 5–15%
- Improve personnel productivity by 7–12%
- Reduce obsolescence and waste by 35–50%

Working capital:

- Reduce inventory levels by 7–15%
- Improve asset utilization by 10–15%
- Decrease cash-to-cash cycle by 10–30%

As it could be seen, there are great benefits on becoming DDSC, but the question that most companies face is ***how to rapidly evolve from current state in the direction of demand-driven supply chain***.

References

Agarwal A, Shankar R, Tiwari M (2007) Modeling agility of supply chain. Ind Mark Manage 36: 443–457

AMR Research Report (2005) The handbook for becoming demand driven. AMR Research, Boston

Ashayeri J, Kampstra RP (2005) Demand driven distribution: The logistical challenges and opportunities. In P.B. Ruffini, J.K. Kim (Eds.), *Proceedings of International Trade & Logistics, Corporate Strategies and the Global Economy*. Le Havre: University of Le Havre.
Ayers J (2006) Handbook of supply chain management, 2nd edn. Auerbach, Boca Raton
Ayers J, Malmberg D (2002) Supply Chain Systems: Are you Ready? Information Strategy: The Executive's Journal
Bayraktar E, Demirbag M, Koh S, Tatoglu E, Zaim H (2009) A causal analysis of the impact of information systems and supply chain management practices on operational performance: evidence from manufacturing SMEs in Turkey. Int J Prod Econ 122(1):133–149
Bowersox D, Lahowchich N (2008) Start pulling your chain: leading responsive supply chain transformation. OGI Enterprises, Port St. Lucie
Childerhouse P, Aitken J, Towill D (2002) Analysis and design of focused demand chains. J Oper Manage 20:675–689
Chin K, Pun K, Lau H (2003) Development of a knowledge-based self assessment system for measuring organizational performance. Expert Syst Appl 24:443–455
Christopher M (2000) The agile supply chain. Ind Mark Manage 29:37–44
Council of Supply Chain Management Professionals – CSCMP (2009) http://cscmp.org/aboutcscmp/definitions.asp. Accessed June 2009
Dale B, Ritchie L (2000) Self-assessment using the business excellence model: a study of practice and process. Int J Prod Econ 66:241–254
Demand Driven Supply Networks: Advancing Supply Chain Management (2006) SAP Insight Report by SAP AG
Emmett S, Crocker B (2006) The relationship driven-supply chain: creating a culture of collaboration throughout the chain. Gower, Aldershot
Filho A, Andrade J, Oliveira M (2010) A safety culture maturity model for petrochemical companies in Brazil. Saf Sci 48:615–624
Fisher M (1997) What is the right supply chain for your product. Harv Bus Rev 75(2):105–116
Georgiadis D, Shinakis M, Tyrinopoulos Y (2001) The design and implementation of a demand driven freight transport application
Harrison T (2003) The practice of supply chain management: where theory and application converge. International series in operations research and management science. Springer, New York
Hopp W, Spearman M (2003) To pull or not to pull: what is the question? Manuf Serv Oper Manage 6(2):133–148
Huang C, Liang W, Lin S (2009) An agile approach for supply chain modeling. Transp Res E 45:380–397
Hull B (2005) Are supply (driven) chains forgotten? Int J Logistics Manage 16(2):218–236
Ismail HS, Sharifi H (2006) A balanced approach to building agile supply chains. Int J Phys Distrib Logistics Manage 36(6):431–444
Longitudes (2007) Competitiveness and the global supply chain. UPS and Harvard Business School, Boston, p 5
Moon M (2002) Establishing best practice: the role of the sales forecasting audit. International Symposium on Forecasting, Dublin
Salama K, Luzzatto D, Sianesi A, Towill D (2009) The value of auditing supply chains. Int J Prod Econ 119:34–45
Taylor D (2000) Demand amplification: has it got us beat? Int J Phys Distrib Logistics Manage 30(6):515–533
Trkman P, McCormack K (2009) Supply chain risk in turbulent environments: a conceptual model for managing supply chain network risk. Int J Prod Econ 119:247–258
Verdouw C, Beulens A, Trienekens J, Wolfert J (2010) Process modeling in demand-driven supply chains: a reference model for the fruit industry. Comput Electron Agric 73(2):174–187
Wanke FP, Correa H, Hijjar M (2010) Establishing the relationship between logistics complexity and supply chain objectives and decision areas in large companies operating in Brazil. J Oper Supply Chain Manage 3(1):34–54
Zeithaml V, Varadarajan P, Zeithaml C (1988) The contingency approach: its foundations and relevance to theory building and research in marketing. Eur J Mark 22(7):37–64

Chapter 3
Proposed Framework

This chapter reviews the research design methodologies and explains how to build theory from case study research. Based on this revision, it is presented the research method used for developing the DDSC framework, and finally, presents the required steps of the Demand Driven Supply Chain assessment framework.

3.1 Briefly Review of Research Design

Yin (2009) defines research design as "*. . .a plan that guides the investigator in the process of collecting, analyzing, and interpreting observations. It is a logical model of proof that allows the researcher to draw inferences concerning causal relations among the variables under investigation. . .*".

He also emphasizes that a research design is much more than a work plan. The main purpose of the design is to avoid situation in which the evidence does not address the initial research questions.

Creswell (2009) also provides a definition of research design as "*. . .plans and procedures for research that span the decisions from broad assumptions to detailed methods of data collection and analysis. . .*"

He presents three types of research designs:

- **Qualitative research** is a means for exploring and understanding the meaning individuals or groups ascribe to a social or human problem. The process of research involves emerging questions and procedures, data typically collected in the participant's setting, data analysis inductively building from particulars to general themes, and the researcher making interpretations of the meaning of the data. The final written report has a flexible structure. The following is a synthesis of commonly articulated assumptions regarding characteristics of a qualitative research:
  - Qualitative research occurs in natural settings, where human behavior and events occur.

P. Mendes, *Demand Driven Supply Chain*,
DOI 10.1007/978-3-642-19992-9_3, © Springer-Verlag Berlin Heidelberg 2011

  - Qualitative research is based on assumptions that are very different from quantitative designs. Theory or hypotheses are not established a priori.
  - The data that emerge from a qualitative study are descriptive. That is, data are reported in words or pictures, rather than in numbers.
  - Qualitative research focuses on the process that is occurring as well as the product or outcome.
  - This research tradition relies on the utilization of tacit knowledge (intuitive and felt knowledge) because often the nuances of the multiple realities can be appreciated most in this way.

- **Quantitative research** is a means for testing objective theories by examining the relationship among variables. These variables, in turn, can be measured, typically on instruments, so that numbered data can be analyzed using statistical procedures. The final written report has a set structure consisting of introduction, literature and theory, methods, results, and discussions. Those who engage in this form of inquiry have assumptions about testing theories deductively, building in protections against bias, controlling for alternative explanations, and being able to generalize and replicate the findings.
- **Mixed methods research** is an approach to inquiry that combines or associates both qualitative and quantitative forms. It involves philosophical assumptions, the use of qualitative and quantitative approaches, and the mixing of both approaches in a study. Thus, it is more than simply collecting and analyzing both kinds of data; it also involves the use of both approaches in tandem so that the overall strength of a study is greater than either qualitative or quantitative research.

Creswell (2009) also states that research design involves the intersection of philosophy, strategies of inquiry and specific methods as illustrated in Fig. 3.1.

- **Philosophical worldviews** – the term "*worldview*" has a meaning of "*a basic set of beliefs that guide action.*" Other authors have called them *paradigms*, *epistemologies* or *broadly conceived research methodologies*. Creswell (2009) describes the worldviews as a general orientation about the world and the nature of research that a researcher holds. The types of beliefs held by individual researchers will often lead to embracing a qualitative, quantitative, or mixed methods approach in their research. Four different worldviews are proposed: Postpositivism, Constructivism, Advocacy and Pragmatism.
- **Strategies of inquiry** – are types of qualitative, quantitative, and mixed methods designs or models that provide specific direction for procedures in a research design. Other authors have called them *approaches to inquiry* or *research methodologies*.

  - **Quantitative strategies** which include surveys and experiments:

    Survey research provides a quantitative or numeric description of trends, attitudes, or opinions of a population by studying a sample of that population. It includes cross-sectional and longitudinal studies using questionnaires or structured interviews for data collection, with the intent of generalizing from a sample to a population.

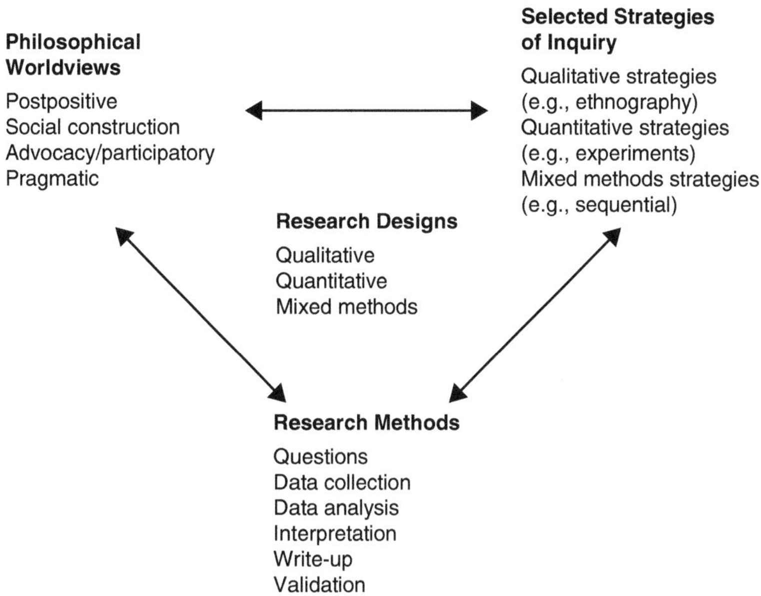

Fig. 3.1 Framework for research design (Creswell 2009)

Experimental research seeks to determine if a specific treatment influences an outcome. This impact is assessed by providing a specific treatment to one group and withholding it from another and then determining how both groups scored on an outcome. Experiments include true experiments, with the random assignment of subjects to treatment conditions, and quasi-experiments that use nonrandomized designs.

– **Qualitative strategies** which include ethnography, grounded theory, case studies, phenomenological research, narrative research.

Ethnography is a strategy of inquiry in which the researcher studies an intact cultural group in a natural setting over a prolonged period of time by collecting, primarily, observational and interview data. The research process is flexible and typically evolves contextually in response to the lived realities encountered in the field setting.

Grounded theory is a strategy of inquiry in which the researcher derives a general, abstract theory of a process, action, or interaction grounded in the views of participants. This process involves using multiple stages of data collection and the refinement and interrelationship of categories of information. Two primary characteristics of this design are the constant comparison of data with emerging categories and theoretical sampling of different groups to maximize the similarities and the differences of information.

Case studies are a strategy of inquiry in which the researcher explores in depth a program, event, activity, process or one of more individuals. Cases are bounded by time and activity, and researchers collect detailed information using a variety of data collection procedures over a sustained period of time.
Phenomenological research is a strategy of inquiry in which the researcher identifies the essence of human experiences about a phenomenon as described by participants. In this process, the researcher brackets or sets aside his or her own experiences in order to understand those of the participants in the study.
Narrative research is a strategy of inquiry in which the researcher studies the lives of individuals and asks one or more individuals to provide stories about their lives. This information is then often retold by the researcher into a narrative chronology.

– **Mixed methods strategies** which include sequential mixed methods, concurrent mixed methods and transformative mixed methods.

Sequential mixed methods procedures are those in which the researcher seeks to elaborate on or expand on the findings of one method with another method. This may involve beginning with a qualitative interview for exploratory purposes and following up with a quantitative, survey method with a large sample so that the researcher can generalize results to a population. Alternatively, the study may begin with a quantitative method in which a theory or concept is tested, followed by a qualitative method involving detailed exploration with a few cases or individuals.
Concurrent mixed methods procedures are those in which the researcher converges or merges quantitative and qualitative data in order to provide a comprehensive analysis of the research problem. In this design, the investigator collects both forms of data at the same time and then integrates the information in the interpretation of the overall results. Also, in this design, the researcher may embed one smaller form of data within another larger data collection in order to analyze different types of questions.
Transformative mixed methods procedures are those in which the researcher uses a theoretical lens as an overarching perspective within a design that contains both quantitative and qualitative data.

- **Research methods** – It involve the forms of data collection, analysis, and interpretation that researchers propose for their studies.

Creswell (2009) proposes to consider the following criteria for selecting a research design:

- **Research problem** – Certain types of social research problems call for specific approaches. For example, problems that call for the identification of factors that influence an outcome, or the utility of an intervention or understanding the best predictors of outcomes, are best suited with quantitative approach. On the other hand, if a concept or phenomenon needs to be understood because little research has been done on it, then a qualitative approach provides a better method.

- **Personal experience** – Researchers' own personal training and experience also influence their choice of approach. For example, individuals trained in technical, scientific writing, statistics, and computer statistical programs will most likely choose the quantitative design. On the other hand, individuals who enjoy writing in a literary way or conducting personal interviews or making up close observations may choose qualitative approach. Qualitative approach allows room to be innovative and to work more within researcher-designed frameworks.
- **Audience** – Researchers write for audiences that will accept their research. These audiences may be journal editors, journal readers, conference attendees, etc. The experiences of these audiences with quantitative, qualitative, or mixed methods studies can shape the decision about this choice.

For more information on the different types of research design and strategies, it is suggested to refer to the work done by Creswell (2009).

3.2 Building Theory from Case Study Research

Yin (2009) defines case study research as "*. . .An empirical inquiry that investigates a contemporary phenomenon in depth and within its real-life context, especially when the boundaries between phenomenon and context are not clearly evident. . .*". He complements this definition saying that case study inquiry:

- Copes with the technically distinctive situation in which there will be many more variables of interest than data points, and as one result relies on multiple sources of evidence, with data needing to converge in a triangulating fashion, and as another result benefits from the prior development of theoretical propositions to guide data collection and analysis.

Yin (2009) states that case study research should be used when a "how" or "why" research question is being asked about a contemporary set of events, over which the researcher has little or no control. He also provides a comparison of case study with other research methods as detailed in Table 3.1 below:

Table 3.1 Relevant situations for different research methods (Yin 2009)

Method	Form of research question	Requires control of behavior events?	Focuses on contemporary events?
Experiment	How, why?	Yes	Yes
Survey	Who, what, where, How many, How much?	No	Yes
Archival analysis	Who, what, where, How many, How much?	No	Yes/No
History	How, why?	No	No
Case study	How, why?	No	Yes

Yin (2009) presents the traditional prejudices against case study method:

- Lack of rigor due to the researcher had not followed systematic procedures, or had allowed equivocal evidence or biased views to influence the direction of the findings and conclusions. For this prejudice, he argues that bias can also enter in the conduct of experiments and the use of research methods, such as designing questionnaires for surveys or conducting historical research.
- Provides little basis for scientific generalization. The answer to this statement is that case studies, like experiments, are generalizable to theoretical propositions and not to populations or universes.
- Cases studies take too long and they result in massive, unreadable documents. This complaint may be appropriate, given the way case studies have been done in the past, but this is not necessarily the way case studies must be done.

Based on Yin (2009), four tests have been commonly used to establish the quality of any empirical social research. Because case studies are one form of such research, the four tests are also relevant to case studies, and are described below:

- Construct validity: Identifying correct operational measures for the concepts being studied.
  - Case study tactics: Use of multiple sources of evidence, establish chain of evidence, have key informants review draft case study report.
- Internal validity: Seeking to establish a causal relationship, whereby certain conditions are believed to lead to other conditions, as distinguished from spurious relationships (for explanatory or causal studies only and not for description or exploratory study).
  - Case study tactics: Pattern matching, explanation building, address rival explanations, use of logic models.
- External validity: Defining the domain to which a study's finding can be generalized.
  - Case study tactics: Use theory in single-case studies, use replication logic in multiple-case studies.
- Reliability: Demonstrating that the operations of a study, such as the data collection procedures can be repeated with the same results.
  - Case study tactics: Use case study protocol, develop case study database.

A primary distinction in designing case studies is between single and multiple case designs. The single-case study is an appropriate design for the following circumstances:

- When it represents the critical case in testing a well-formulated theory. The theory has specified a clear set of propositions as well as the circumstances within which the propositions are believed to be true, and the single case that meets all of the conditions for testing the theory, can confirm, challenge, or extend the theory.

- When the case represents an extreme case or a unique case.
- When a single case is the representative or typical case. In this case, the objective is to capture the circumstances and conditions of an everyday situation.
- When it is a revelatory case. This situation exists when an investigator has an opportunity to observe and analyze a phenomenon previously inaccessible to social science inquiry.
- When it is a longitudinal case, which means, studying the same single case at two or more different points in time.

The same study may contain more than a single case. When this occurs, the study has used a multiple-case design, and such designs have increased in frequency in recent years.

Based on Yin (2009), multiple-case designs have distinct advantages and disadvantages in comparison to single-case designs. The evidence from multiple cases is often considered more compelling, and the overall study is therefore regarded as being more robust (Herriott and Firestone 1983). At the same time, the rationale for single-case designs cannot usually be satisfied by multiple cases when it is considered unusual or rare case, critical case, and the revelatory case.

He also states that the replication logic for multiple-case studies is analogous to that used in multiple experiments. Some of the replications might attempt to duplicate the exact conditions of the original experiment while others might alter one or two experimental conditions considered unimportant to the original finding, to confirm whether the finding could still be duplicated. Each case must be carefully selected so that either predicts similar results (literal replication) or predicts contrasting results but for anticipatable reasons (theoretical replication). A few cases (2 or 3) would be literal replications, whereas a few other cases (4–6) might be designed to pursue two different patterns of theoretical replications.

Eisenhardt (1989) describes the process of inducting theory using case studies from specifying the research questions to reaching closure. She argues that this research approach is especially appropriate in new topic areas, and the resultant theory is often novel, testable, and empirically valid.

She states that in normal science, theory is developed through incremental empirical testing and extension, and thus, the theory-building process relies on past literature and empirical observation or experience as well as on the insight of the theorist to build incrementally more powerful theories. However, she highlights that there are times when little is known about a phenomenon, current perspectives seem inadequate because they have little empirical substantiation, or they conflict with each other or common sense, or there is a need for a new perspective. In these situations, theory building from case study research is particularly appropriate because theory building from case studies does not rely on previous literature or prior empirical evidence.

Table 3.2 Strengths and weaknesses of building theory from case study (Eisenhardt 1989)

Strengths	Weaknesses
Likelihood of generating novel theory	Intensive use of empirical evidence can yield theory which is overly complex
Resultant theory is likely to be empirically valid. The likelihood of valid theory is high because the theory-building process is so intimately tied with evidence that it is very likely that the resultant theory will be consistent with empirical observation	Building theory from cases may result in narrow and idiosyncratic theory. It is a bottom up approach and there is a risk of not being able to raise the level of generality of the theory
Emergent theory is likely to be testable with constructs that can be readily measured and hypotheses that can be proven false	

In sum, building theory from case study research is most appropriate in the early stages of research on a topic or to provide freshness in perspective to an already researched topic.

These characteristics of theory building from case study seem to fit well with the proposed problem stated in this book as the main objective is to develop a framework for assessing and guiding companies' progress towards a Demand Driven Supply Chain concept, which is not clearly defined yet in the academic literature and not fully executed in practice by supply chain professionals.

Before establishing this method as the basis for this research, it is important to understand the strengths and weaknesses of theory-building from case study, which is summarized in Table 3.2 above:

In terms of evaluation of theory-building research using case studies, Eisenhardt (1989) states that there is no generally accepted set of guidelines for the assessment of this type of research, but several criteria seem appropriate like:

- Assessment on whether the concepts, framework, or propositions that emerge from the process are "good theory," which is defined as parsimonious, testable, and logically coherent theory.
- Assessment of theory building based on the strength of method and the evidence of grounding theory like checking if the researchers followed a careful analytical procedure, that the evidence support the theory.
- Strong theory-building research should result in new insights. Theory building which simply replicates past theory is, at best, a modest contribution. Thus, a strong theory building study presents new, perhaps framebreaking insights.

Another important contribution to develop solutions to practical problems that follows the same "*school of thought*" comes from Van Aken (2004), where he proposes the paradigm of *Design Sciences* which has the mission to develop knowledge to solve construction problems, or to be used in the improvement of the performance of existing entities, i.e., to solve improvement problems.

He explains the differences of design sciences to *formal sciences* which has the mission to build systems of propositions whose main test is their internal logical

Table 3.3 Main differences between explanatory and design sciences (Van Aken 2004)

Characteristic	Description-driven research programmes	Prescription-driven research programmes
Dominant paradigm	Explanatory sciences	Design sciences
Focus	Problem focused	Solution focused
Perspective	Observer	Player
Logic	Hindsight	Intervention-outcome
Typical research question	Explanation	Alternative solutions for a class of problems
Typical research product	Causal model; quantitative law	Tested and grounded technological rule
Nature of research product	Algorithm	Heuristic
Justification	Proof	Saturated evidence
Type of resulting theory	Organization theory	Management theory

consistency, and also to *explanatory sciences* which has the mission to describe, explain and possibly predict observable phenomena within its field. Table 3.3 above summarizes the main differences between explanatory sciences and design sciences:

A design science does not develop knowledge for the layman, but rather for professionals in its field, which means that design knowledge is to be applied by individuals who have received formal education in that field. In the design sciences, the research object is a "*mutandum,*" and these sciences are not much interested in *what is*, but more in *what can be*. The typical research product is a "*technological rule*" which is defined by Van Aken as "*...a chunck of general knowledge, linking an intervention or artifact with a desired outcome or performance in a certain field of application...*"

He also emphasize that professionals have a repertoire of design knowledge at their disposal to make these design, which one of them is their own personal experience.

For more information about the design sciences, please refer to Van Aken (2004).

3.3 Research Method for Developing DDSC Assessment Framework

The research followed a design-oriented methodology similar to the one applied by Verdouw et al. (2010) to develop a reference process modeling for the fruit industry in Europe to become a demand driven supply chain.

Design-oriented research is typically involved with "how" questions, i.e., how to design a model or system that solves a certain problem, as stated by Van Aken

(2004). This research applies a design-testing approach, which is comparable with theory testing methods in traditional empirical science, as explained by Eisenhardt (1989). In such approach, generic design knowledge is developed based on deductive reasoning, and after that, the design is tested by applying it to specific cases. In this research, the proposed assessment framework is applied in a multiple case study in a CPG global company.

Four different countries were preselected to review the proposed maturity model and answer the assessment to identify current and future states based on demand driven supply chain concepts. In this way, it will be possible to validate the proposed maturity model at the same time that companies' current and future states are identified. The main criteria to select the countries were their market maturity, interviewees' supply chain practical experience, and author's knowledge of their operations.

The research is organized in four main steps (1) Literature review, (2) Development of maturity model, (3) Case investigation and analysis through application in four operations and (4) Review of maturity model based on feedback from practitioners. A proposed framework describing each step was developed and is detailed in the next section.

3.4 Proposed DDSC Framework

In this section, the proposed framework used to develop the DDSC maturity model will be described, and also the proposed steps to perform the assessment process.

A methodology is defined as a structured collection of stages that have to be carried out in order to achieve business improvement. This is usually given in the form of a flowchart that defines, for each stage, what should be done, when, how, why and by whom.

In order to solve the problem previously described, it is proposed a new framework consisting of a two-phase approach that is illustrated in the framework of Fig. 3.2.

The first phase is called the "**Construction Phase**" and aims to identify the DDSC components and develop the Demand Driven Supply Chain Maturity Model, and comprehends three major steps, as described below:

- **Steps 1 and 2 – Literature Review and Identify DDSC Component:**

In the first two steps, the academic literature currently available is reviewed and also explored practical experience from the author, who has more than 17 years of practical experience in leading and developing logistics and supply chain management projects worldwide to identify and define the components of a Demand Driven Supply Chain.

This is a very important step in the proposed methodology as the DDSC concepts are not currently gathered and documented in one single source, as confirmed during the literature review done in Chap. 2.

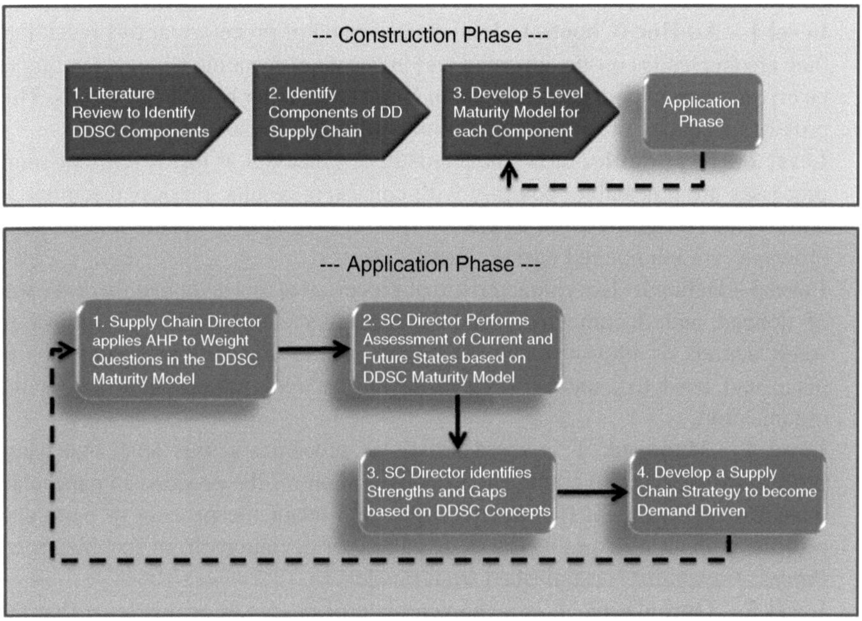

Fig. 3.2 Author's integrated methodology to assess DDSC

- **Step 3 – Develop Five-Level Maturity Model for Each Component:**

Based on the characteristics identified in the previous steps, a five level maturity model will be developed, ranging from a level 1 (low adherence) to level 5 (full implemented) of DDSC concepts that will serve as the basis to perform the assessment of current and future states.

A maturity model can be described as a structured collection of elements that describe certain aspects of maturity in an organization, and aids in the definition and understanding of the different organization processes.

The Capability Maturity Model (CMM) was originally developed as a tool for objectively assessing the ability of government contractors' processes to perform a contracted software project. The CMM is based on the Process Maturity Framework first described in "Managing the Software Process" by Watts Humphrey and later published in its full form as a book named The Capability Maturity Model: Guidelines for Improving the Software Process in 1994 by Mark Paulk, Charles Weber, Bill Curtis, and Mary Beth Chrissis.

Though it comes from the area of software development, it has also been applied to improving organizational processes in diverse areas, like software engineering, system engineering, project management, software maintenance, risk management, system acquisition, information technology (IT), services, business processes, and human capital management.

There are five standard levels defined along the continuum of the CMM, as described below:

Level 1 – Ad Hoc (Chaotic). It is characteristic of processes at this level, that they are (typically) un-documented and in a state of dynamic change, tending to be driven in an ad hoc, uncontrolled and reactive manner by users or events. This provides a chaotic or unstable environment for the processes.
Level 2 – Repeatable. It is characteristic of processes at this level, that some processes are repeatable, possibly with consistent results. Process discipline is unlikely to be rigorous, but where it exists, it may help to ensure that existing processes are maintained during times of stress.
Level 3 – Defined. It is characteristic of processes at this level that there are sets of defined and documented standard processes established, and subject to some degree of improvement over time. These standard processes are in place and used to establish consistency of process performance across the organization.
Level 4 – Managed. It is characteristic of processes at this level that using process metrics, management can effectively control the process. In particular, management can identify ways to adjust and adapt the process to particular projects without measurable losses of quality or deviations from specifications. Process Capability is established from this level.
Level 5 – Optimizing. It is a characteristic of processes at this level that the focus is on continuous improving process performance through both incremental and innovative technological changes/improvements.

The second phase is called an "**Application Phase**" and aims to apply the framework in the different operations and countries to identify the current state and develop the supply chain strategy to become a Demand Driven organization, and comprehends 4 major steps, described below:

- **Step 1 – Supply Chain Director Applies AHP to Weight Components and Categories in the DDSC Maturity Model:**

  The first step of the Application Phase consists of having the supply chain directors, who are responsible to develop the supply chain strategy for their organizations, providing their view and weighting the importance for each component and categories in the DDSC maturity model through a set of different weights that needs to be reconciled in one integrated view for the company under analysis.
  The application of an audit/assessment process is referred by Salama et al. (2009) as an important step companies should perform in order to achieve business improvements and face the competitive pressure of today's high dynamic markets. They also argue that sometimes process related problems are not solved because companies fail to identify them, and on the other hand, the evaluation of innovative technologies or managerial practices can represent a way not only to solve hidden problems, but also to develop new business models and allow to do things that the organization is not already doing.

Table 3.4 Advantages and disadvantages of AHP

Advantages	Disadvantages
Accommodates multiple criteria	Length of the process, which increases with the number of levels and number of pairwise comparisons
Model is simple, intuitive, but has mathematical rigor	Expense of commercial software to make the approach easily implemented
Integrates subjective judgments with numerical data	If not properly implemented, it can generate inconsistencies due to the pairwise comparisons
Facilitate decision maker participation	
Encourages a process of learning, debate and revision from multiple participants	
Allows building alternative scenarios to cope with medium/long term uncertainty	

To that end, it is needed a robust approach to generate this final integrated set of weights, and in this methodology, it is proposed to apply the Analytic Hierarchic Process (AHP).

Based on Saaty and Vargas (2006) there are two known ways to analyze causal influences and their effects. One is by using traditional deductive logic beginning with assumptions and carefully deducing an outcome from them. This is a linear and piecemeal approach in which several separate conclusions may be obtained and the problem is to piece them together in some coherent way to have an integrated outcome.

The other way is to have a holistic approach in which all the factors and criteria involved are laid out in advance in a hierarchy or in a network system that allows for dependencies. All possible outcomes that can be thought of are joined together in these structures, and then, both judgment and logic are used to estimate the relative influence from which the overall answer is derived. This approach requires knowledge and experience with the subject, and is not totally dependent on the ability to reason logically which most people cannot do well.

Table 3.4 above provides a summary of the advantages and disadvantages of the AHP process:

The AHP is a general theory of measurement. It is used to derive relative priorities on absolute scales from both discrete and continuous paired comparison in multilevel hierarchic structures. The AHP has a special concern with departure from consistency and the measurement of this departure, and with dependence within and between the groups of elements of its structure. In order to use the AHP to model a problem, a hierarchic structure to represent the problem is needed, as well as pairwise comparisons to establish relations within the structure.

- **Step 2 – Supply Chain Director Performs Assessment of Current and Future States Based on DDSC Maturity Model:**

The second step is to have supply chain directors performing an assessment of current and future states in light of the DDSC concepts.

This step should be performed very honestly and with an open mind in order that the results really reflect the current state of the operation under analysis, otherwise, the company will not be able to understand what the improvement opportunities are and how to move towards DDSC.

- **Step 3 – Identify Strengths and Gaps Based on DDSC Concepts:**

  Based on the results, the company will be able to identify the strengths and gaps of current state, and use this information to develop a supply chain strategy to move to future state, which represents the desired state in one year time in the future.

- **Step 4 – Develop a Supply Chain Strategy to Become DDSC:**

  The last step of the framework is to develop a supply chain strategy that will allow the company to identify the steps required to become a demand driven supply chain. This development should be performed aligned with the company strategic business planning process, as supply chain is a key enabler of business improvement and can help the company achieve top level business goals like revenue growth, increase asset utilization and profitability, improve customer service, just to name a few examples.

In order to have formal evidence that the proposed framework is robust, methodological consistent and practical, supply chain directors for the first implementation will be asked to provide their feedback about the five-level maturity model descriptions and the proposed approach using AHP model.

After applying the framework in practice and based on the feedback received from supply chain directors, the maturity model and the proposed approach will be reviewed if needed, to make any necessary adjustment or changes, in order to better reflect the concepts of demand driven supply chain.

References

Creswell J (2009) Research design: qualitative, quantitative, and mixed methods approaches, 3rd edn. Sage, Thousand Oaks

Eisenhardt K (1989) Building theory from case study research. Acad Manage Rev 14(4):532–550

Herriott R, Firestone W (1983) Multisite qualitative policy research: optimizing description and generalizability. Educ Res 12:14–19

Paulk M, Weber C, Curtis B, Chrissis M (1994) The Capability Maturity Model: Guidelines for Improving the Software Process. Addison-Wesley Professional

Saaty T, Vargas L (2006) Decision making with the analytic network process: economic, political, social and technological applications with benefits, opportunities, costs and risks, 1st edn. Springer, New York

Salama K, Luzzatto D, Sianesi A, Towill D (2009) The value of auditing supply chains. Int J Prod Econ 119:34–45

Van Aken J (2004) Management research based on the paradigm of the design sciences: the quest for field-tested and grounded technological rules. J Manage Stud 41:219–246

Verdouw C, Beulens A, Trienekens J, Wolfert J (2010) Process modeling in demand-driven supply chains: a reference model for the fruit industry. Comput Electron Agric 73(2):174–187

Yin R (2009) Case study research: design and methods, 4th edn. Sage, Thousand Oaks

Chapter 4
Key Components of Demand Driven Supply Chain

In this chapter, it will be briefly reviewed the supply chain processes based on the work developed by Lambert (2008), then it will be presented the three Demand Driven Supply Chain components proposed by the author, followed by a literature review for each one of the components. The reason to choose Lambert's processes review was due to its broad coverage of supply chain processes, including customers and suppliers.

4.1 Supply Chain Processes

Lambert (2008) states that empirical research has led to the conclusion that the structure of activities within and between companies is a critical cornerstone of creating unique and superior supply chain performance, and therefore, corporate success requires a change from managing individual functions to integrating activities into supply chain management processes.

He also emphasizes the importance of having standard business processes across the members of the supply chain, in order to have a "common language" that allows integrating processes from different companies.

Lambert (2008) describes that the Global Supply Chain Forum proposes the framework presented in the Fig. 4.1 to integrate and manage business processes across the supply chain.

This framework presents eight key processes that are common to all companies in a supply chain, as summarized below:

- **Customer relationship management** provides the structure for how the relationship with customers will be developed and maintained. The goal is to segment customers based on their value over time and increase customer loyalty of target customers by providing customized products and services. Cross-functional customer teams tailor Product and Service Agreements (PSA) to meet the needs of key accounts and other business segments, and also work

DOI 10.1007/978-3-642-19992-9_4,

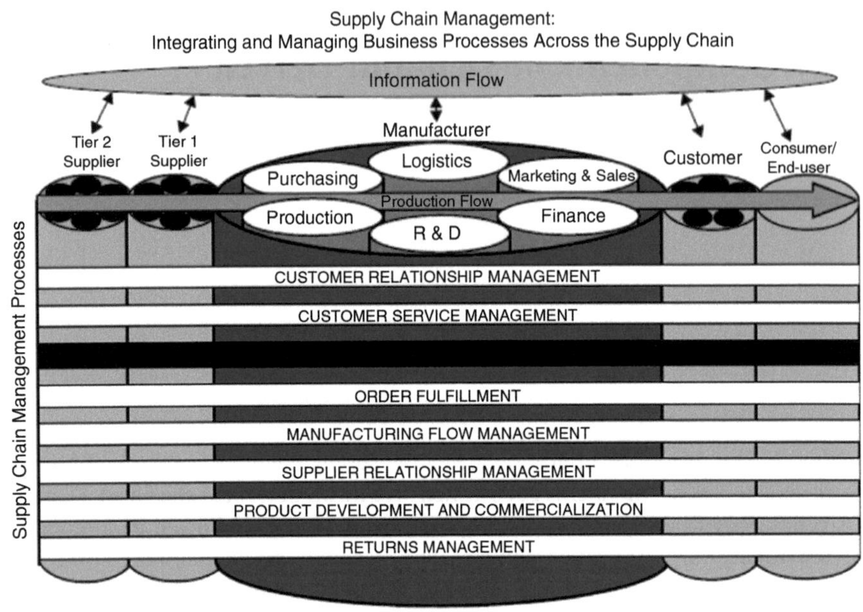

Fig. 4.1 Supply chain processes (Lambert 2008)

together to improve processes and reduce demand variability and non-value added activities.

- **Customer service management** is the process that deals with the administration of the PSAs developed by customer teams as part of the customer relationship management process. Customer service managers monitor the PSAs and proactively intervene on the customer's behalf if there is going to be a problem to deliver the promise that has been made. The goal is to solve problems before they affect the customers.
- **Demand management** is the process that balances the customers' requirements with the capabilities of the supply chain. With the right process in place, management can match supply with demand proactively and execute the plan with minimal disruptions. A good demand management process uses point-of-sale and key customer data to reduce uncertainty and provide efficient flows throughout the supply chain.
- **Order fulfillment** includes all activities necessary to design a network and enable the firm to meet customer demand while minimizing the total delivery cost. In this case, much of the actual work will be performed by the logistics function, but it also requires coordination with key suppliers and customers.
- **Manufacturing flow management** is the process that includes all activities necessary to obtain, implement and manage manufacturing flexibility in the supply chain to move products into, through and out of the plants. Manufacturing flexibility reflects the ability to make a wide variety of products in a timely manner at the lowest possible cost.

- **Supplier relationship management** provides the structure for how relationships with suppliers will be developed and maintained. Close relationships are developed with a small subset of suppliers based on the value that they provide to the organization, and more traditional relationships are maintained with others. Supplier teams negotiate PSAs with each key supplier, and for less critical suppliers, a standard PSA is provided and it is not negotiable.
- **Product development and commercialization** is the process that provides the structure for developing and bringing to market products jointly with customers and suppliers. Effective implementation of the process not only enables management to coordinate the efficient flow of new products across the supply chain, but also assists other members of the supply chain with the ramp-up of manufacturing, logistics, marketing, and other activities necessary to support the commercialization of the product.
- **Returns management** is the process by which activities associated with returns, reverse logistics, gate keeping, and avoidance are managed within the firm and across key members of the supply chain. The correct implementation of this process enables management not only manages the reverse product flow efficiently, but also identifies opportunities to reduce unwanted returns and control reusable assets, such as containers, empty bottles, etc.

4.2 Components of Demand Driven Supply Chain

Table 4.1 compares the characteristics of a Demand Driven Supply Chain described in the literature review presented in Chap. 2 with the Supply Chain Processes described by Lambert (2008) in Sect. 4.1, and it can be seen that the supply chain processes represent a possible way to categorize the DDSC components.

For the sake of simplicity and due to the interrelations of some processes, like customer relationship and customer service management, the author proposes to categorize the eight processes from Lambert (2008) in three key components, as explained below:

- **Demand Management,** which encompasses all aspects related to reading, sensing, shaping and synchronizing customer demand.
- **Supply and Operations Management,** which covers all aspects of providing right product with right quantity, based on actual demand signal and with low cost. This component should encompass:
  - **Procurement (Supplier Relationship Management).**
  - **Manufacturing Flow Management.**
  - Order Fulfillment and Return Management, which will be considered as **Logistics Management.**
  - Customer Service Management and Customer Relationship Management, which will be considered as **Sales Management.**

- **Product Lifecycle Management**, which comprehends new product introduction and product sunset, in order to reduce supply chain complexity and allow becoming agile.

Table 4.1 Author's comparison of DDSC characteristics and SC processes

Demand driven SC characteristics	Supply chain process
Capacity to sense and respond to real time demand across the supply chain	Demand management
The higher the demand uncertainty, the more the firm should prefer to manage based on actual demand	Demand management
Customer activates the replenishment flow in the supply chain	Order fulfillment and demand management
Being market driven, understanding customers and markets	Customer relationship management and customer service management
Have a demand driven replenishment process (Pull based system)	Order fulfillment
Lean manufacturing	Manufacturing flow management
Develop products that drives demand, getting excellence in innovation	Customer relationship management and product lifecycle management
DDSC is the foundation of lean supply chain, allowing to operate in a low cost production environment	Manufacturing flow management
Being agile, capable of respond rapidly to changes in demand, both in terms of volume and variety	Demand management and Product lifecycle management
Reduce complexity – product and brand proliferation, organization structure and management process	Product lifecycle management
Build agile networks to align materials suppliers, contract manufacturers, and logistics providers to a demand signal	Supplier management
Agile networks start with the design and flexibility based on joint agreements (contract relationship and demand visibility are essential)	Customer relationship management and supplier relationship management

It is important to mention that each one of the three components above will be detailed and characterized in the sections below, and based on these characteristics and the author experience, a Demand Driven Maturity Model will be developed and used to assess the current state of organizations in light of Demand Driven concepts.

4.3 Demand Management

In this section, it will be performed a literature review for each one of the four categories of the Demand management – Statistical Forecast, Sales and Operations Planning (S&OP), Collaborative Planning and Forecasting Replenishment (CPFR) and Vendor managed Inventory (VMI). This review allowed identify the DDSC characteristics for each category which was used to develop the five level maturity model.

Based on Croxton et al. (2002), the demand management process is concerned with balancing the customers' requirements with the supply chain capabilities. This includes forecasting demand and synchronizing it with production, procurement, and distribution capabilities. A good demand management process can enable a company to be more proactive to anticipated demand.

An important component of demand management is finding ways to reduce demand variability and improve operational flexibility. They argue that reducing demand variability aids in consistent planning and reduce costs, and that increasing flexibility helps the firm respond quickly to internal and external events.

Most customer-driven variability is unavoidable, but one of the goals of demand management is to eliminate management practices that create noise and increase variability, and to introduce policies that foster smooth demand patterns. Another key part of demand management is developing and executing contingency plans when there are interruptions to the operational plans. The goal of demand management is to meet customer demand in the most effective and efficient way.

The demand management process can have a significant impact on the profitability of a firm, its customers and suppliers. Some examples are:

- Having the right product on the shelves will increase sales and customer loyalty.
- Improved forecasting can reduce raw materials and finished goods inventories.
- Smoother operational execution will reduce logistics costs and improve asset utilization.

For Croxton et al. (2002), demand management is about forecasting and synchronizing, and has both strategic and operational sub-processes, as shown in Fig. 4.2.

When Croxton et al. (2002) detailed the sub-process of "determine forecasting procedures," they explain that the first step is to understand what type of forecast is needed, then what data is available, and finally, select a forecasting method which will depend on the environment that the forecasting is taking place. They presented a two-by-two matrix to show which forecast approach is appropriate based on demand variability and demand volume, as shown in Fig. 4.3.

This matrix shows that products with high variability and high volume require more human input from sales or customers, as the statistical quantitative methods alone will not be able to provide good forecast accuracy.

The second case is when a product has low volume and high variability, which in this case a make-to-order production strategy (or pull system) should be used, which avoids the need for an SKU level forecast.

The last case is when a product has low demand variability, and in this case, a data driven statistical forecast should be applied, as it will allow capture the benefits of a push system. The approach described above brings light to help define when a company should be "demand driven" or "forecast driven." Based on Croxton et al. (2002), it is proposed to expand the matrix to also include the tools and approaches that can be used in each one of the three situations, as detailed and illustrated in Fig. 4.4.

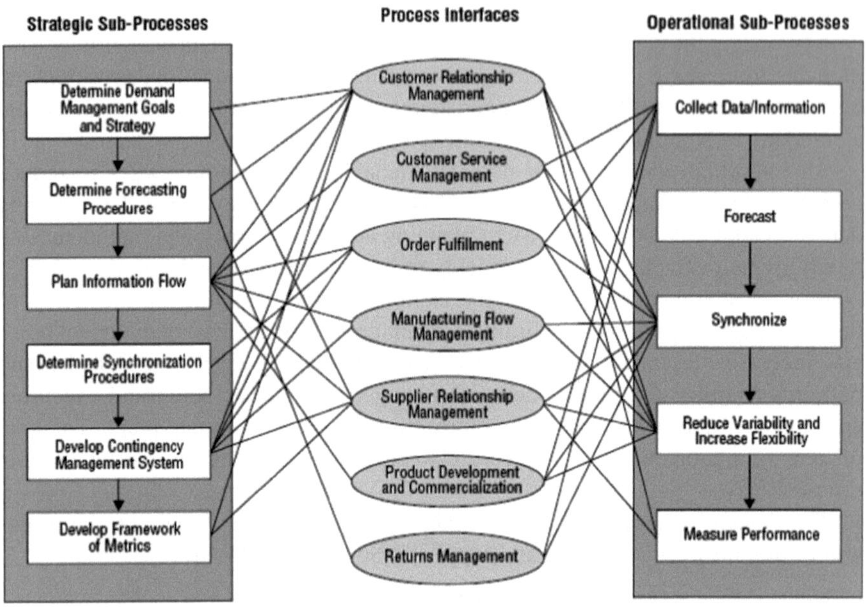

Fig. 4.2 Demand management framework based on Croxton et al. (2002)

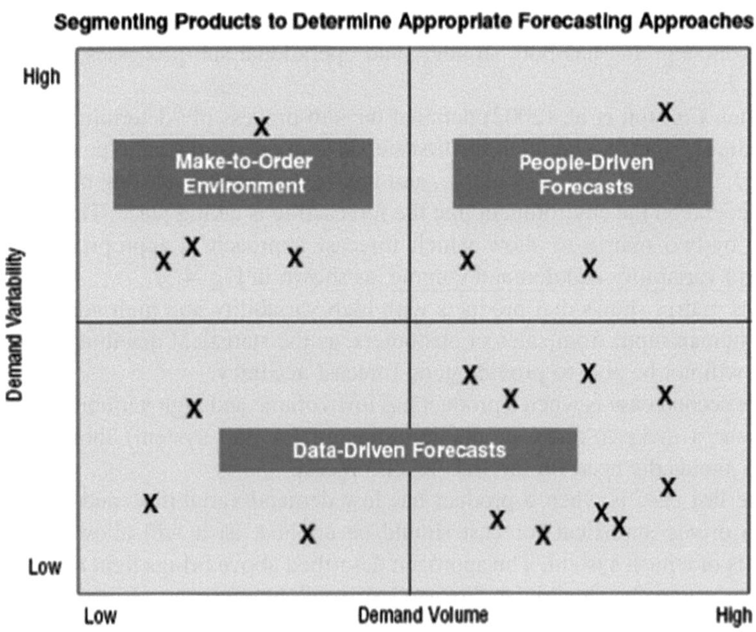

Fig. 4.3 Two by two matrix (Croxton et al. 2002)

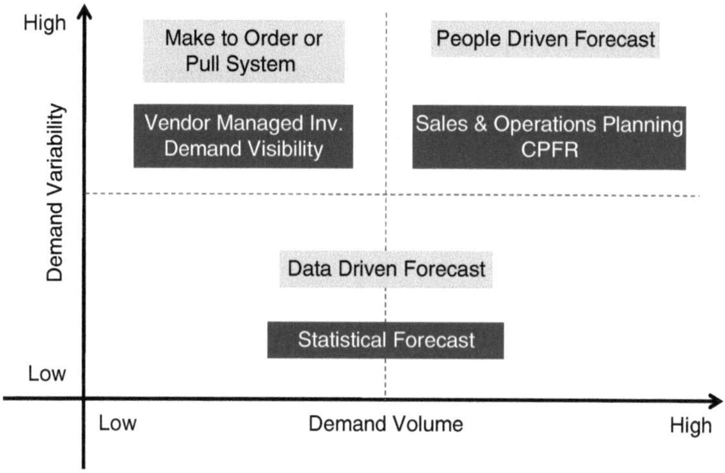

Fig. 4.4 Author's proposed expanded 2 × 2 matrix

- For "data driven forecast," it is suggested to apply statistical forecast models, which will generate good forecast accuracy results, and will also automate the forecasting calculation, saving demand planners' time to devote to more complicated and/or variable SKUs.
- For "make to order," it is suggested to apply VMI and demand visibility to sense the demand signals and quickly react to fulfill it.
- For "people driven forecast," it is suggested to apply Sales and Operations Planning (S&OP) and Collaborative Planning and Forecast Replenishment (CPFR), as they represent structured and formal processes to align demand figures across different functional areas and different organizations.

Each one of the proposed tools and processes showed in Fig. 4.4 will be described in detail in the next sections.

4.3.1 Statistical Forecast

4.3.1.1 Introduction

In management and administrative situations, the need for planning is great because the lead time (time lag between awareness of an event and the occurrence of that event) for decision making ranges from several years (for the case of capital investments) to a few days or hours (for transportation or production scheduling) to a few seconds (for telecommunication routing), and forecasting is an important aid in effective and efficient planning.

The trend to be able to more accurately predict a wider variety of events, particularly those in the economic/business environment, will continue to provide a better base from which to plan, and formal forecasting methods are the means by which this improvement is possible.

Makridakis (1998) states that a wide variety of forecasting methods are available to management and range from the most naïve methods to highly complex approaches, such as neural nets and econometric systems of simultaneous equations.

Chatfield (2004) pointed out that it is also important to realize that no single method is universally applicable. Rather, the analyst must choose the procedure that is most appropriate for a given set of conditions. Forecasts are conditional statements about the future based on specific assumptions, and thus, forecasts are not sacred and the analyst should always be prepared to modify them as necessary in the light of any external information. For long-term forecasting, it can be helpful to produce a range of forecasts based on different sets of assumptions, so that, alternative "scenarios" can be explored.

Moon et al. (1998) state that a sales forecasting is a management function that companies often fail to recognize as a key contributor to corporate success. From a top-line perspective, accurate sales forecasts allow a company to provide high levels of customer service, delivering volume in a timely and efficient manner, keeping both channel partners and final customers satisfied. Accurate forecasts help a company avoid lost sales or stock-out situations, and prevent customers from going to competitors. At the bottom line, the effect of accurate forecasts can be profound. Raw materials and component parts can be purchased much more cost-effectively, logistical services can be obtained at a lower cost and inventory levels can be reduced.

In order to get senior management support to develop the demand planning function inside the organization is critical to know how to measure the impact of forecast error on the company finance results. Mentzer (1999) states that senior levels are not concerned about the forecast accuracy results, but for the impact that improved forecasting accuracy can have on shareholder value. To that end, he proposed to use the "Du Pont Model" of financial performance. (The Du Pont Model is a framework for viewing the impact of changes in sales, capital, and operating expenses on return on net assets.) Figure 4.5 shows an example of the Du Pont Model.

Another key successful factor in forecasting is to have the right organizational structure, which means, the right number of people, with the right skills and performance metrics, and with the right report level.

Lapide (2003) argues that a combination of skills, organization and corporate culture form the cocktail that drives the success of any demand planning department. He describes the following critical skills:

- Quantitative skills because forecasting often involves the use of statistical forecasting methods and algorithms.
- Computer skills, since demand forecasts are often need to be done at the Stock Keeping Unit (SKU) level for multiple periods of time, which usually generate tens or hundreds of thousands of entities.

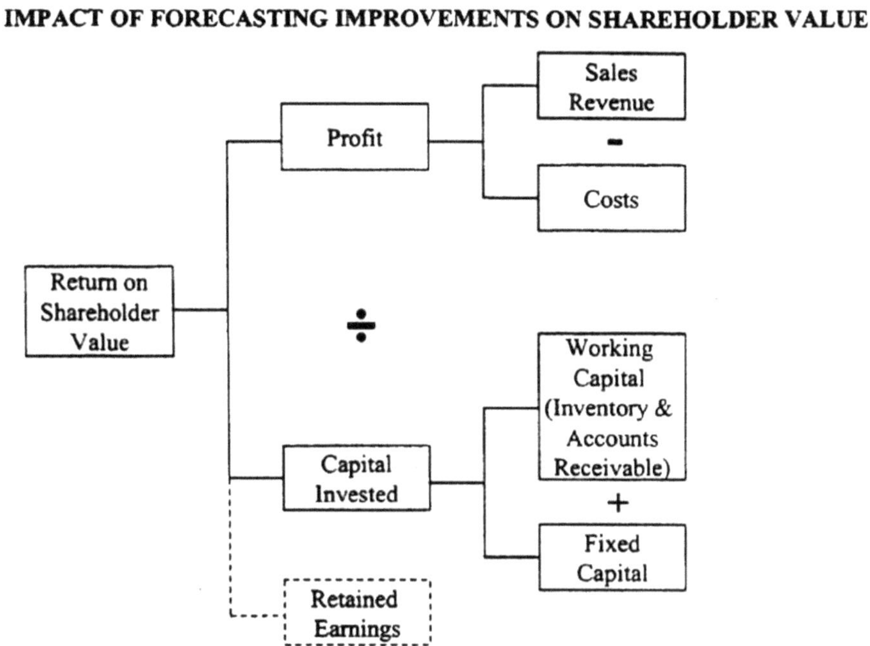

Fig. 4.5 Du Pont model proposed by Mentzer (1999)

- Interpersonal skills to be able to communicate with other departments in order to gather the market intelligence needed to develop and adjust the baseline forecast.
- Understand the business in order to read market signals and identify demand variations.
- Process management skills to ensure to get a one number forecast. This entails getting a cross-functional team to come to a consensus on the forecasts.

Mentzer and Cox (1984) reported a study to analyze the corporate and forecast factors which affect forecast accuracy. The research revealed that the most important corporate factor was formal training of forecast personnel, which means more formal training received the greater achieved forecast accuracy.

Mentzer and Davis (2007) propose a theory-based sales forecasting management (SFM) framework, as illustrated in Fig. 4.6, consisting of four components (sales forecasting climate, capability, performance outcomes, and performance measurement), to facilitate the exploration of the effects of organizational factors in sales forecasting. They argue that a firm's sales forecasting climate influences its sales forecasting capability, which in turn determines performance outcomes.

Mentzer and Moon (2005) propose to have "multidimensional metrics," in other words, metrics that cover three dimensions to sales forecasting performance, in

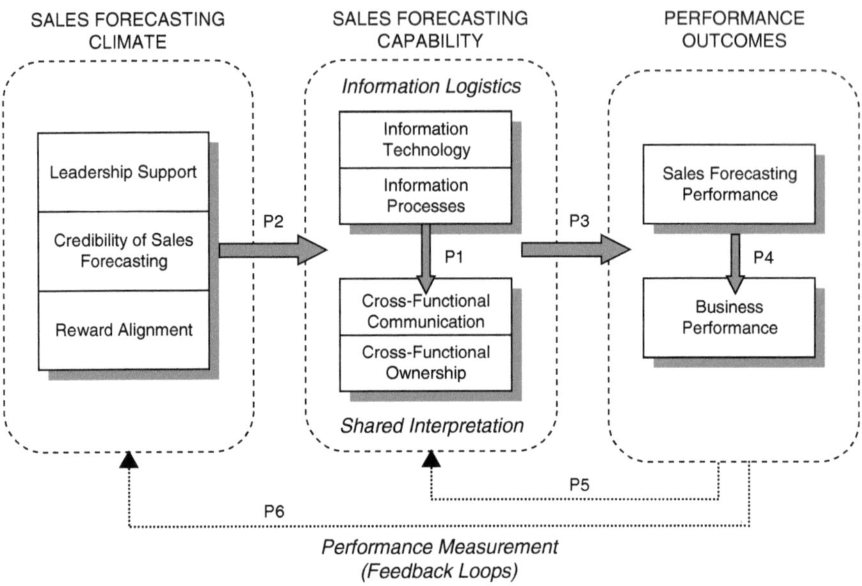

Fig. 4.6 Sales forecasting management proposed by Mentzer and Davis (2007)

order to clear define for those responsible for sales forecasting what needs to be improved. Each of the three dimensions is described below:

1. Accuracy, which comprehends actual measures like the Mean Error (ME), Mean Absolute Error (MAE), Mean Squared Error (MSE), and Measures relative to a Perfect Forecast like the Percent Error (PE) and the Mean Absolute Percent Error (MAPE).
2. Costs, which comprehend operations costs, related to production and logistics, and marketing costs, related to trade promotions, ineffective advertising, and product development without adequate demand, wrong pricing that does not maximize profit contribution.
3. Customer Satisfaction which comprehends customer surveys to determine their satisfaction with all company activities. It should cover customers' perception of the timeliness, availability and condition of the distribution service they receive, and finally, the overall customer satisfaction process.

Taylor (2000) developed an approach to eliminate demand amplification. The approach developed had the objective to be sufficiently straightforward to be easily applied by staff that has day-by-day responsibility for managing demand along a supply chain. There are seven steps to be followed:

- Identify and quantify demand amplification
- Analyze the specific causes of the effect in the supply chain under study
- Education and awareness raising with relevant personnel
- Creation of a demand management team from across the supply chain

- Development and application of detailed policies to address the effect in selected trial value streams
- Monitoring and evaluation of supply chain performance during the trial
- Roll out the policy to other value streams, modifying in light of the trial

4.3.1.2 Forecast Models

Chatfield (2004) states that forecasting methods can be broadly classified into three group as follows:

1. **Subjective:** Forecasts can be made on a subjective basis using judgment, intuition, commercial knowledge or any other relevant information. Methods range widely, like the Delphi technique, in which a group of forecasters tries to obtain a consensus forecast with controlled feedback of other analysts' predictions and opinions, and other relevant information.
2. **Univariate**: Forecasts of a given variable demand are based on a model fitted only to present and past observations of a given time series. There are several different univariate models, like Extrapolation of Trend Curves, Simple Exponential Smoothing, Holt Method, Holt–Winters Method, Box-Jenkins Procedure, and Stepwise Auto-regression, which can be regarded as a subset of the Box–Jenkins Procedure.
3. **Multivariate:** Forecasts of a given variable depend at least partly on values of one or more additional variables, called explanatory variables. Models of this type are usually called "causal models," and include Multiple Regression, and Econometric Models.

Makridakis (1998) classifies the forecasting techniques into two categories:

1. **Quantitative** to be considered when sufficient quantitative information is available. In this case, it can be sub-divided into:
   (a) Time Series, which is a collection of observations made sequentially through time. In this class of models, prediction of the future is based on past values of a variable and/or past error, but not on explanatory variables.
   (b) Explanatory, which assumes that the variable to be forecasted exhibits an explanatory relationship with one or more independent variables.
2. **Qualitative** to be considered when little or no quantitative information is available, but sufficient qualitative knowledge exists.

He also states that quantitative forecasting can be applied when three conditions exist:

- Information about the past is available
- This information can be quantified in the form of numerical data
- It can be assumed that some aspects of the past pattern will continue into the future

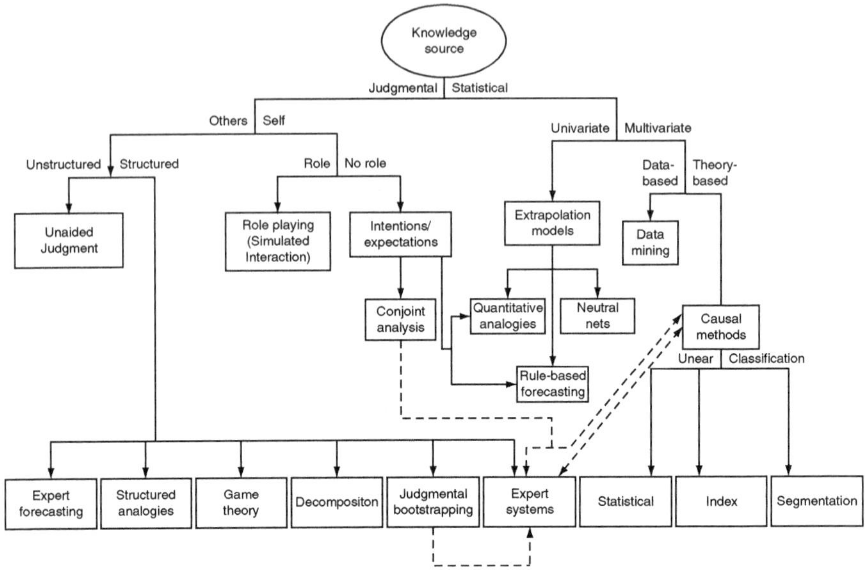

Fig. 4.7 Methodology tree proposed by Armstrong (2001)

Armstrong (2001) proposes a methodology tree to classify forecast models in different categories as detailed in Fig. 4.7.

It is not the objective of this thesis to review in detail each one of the forecast models currently available, but instead, list the key models that should be used in forecasting for business and operations. For a detailed explanation of each forecast model, please refer to the work developed by Mentzer and Moon (2005), Makridakis (1998) or Oliveira Junior (2004).

Regarding the steps required to perform a statistical forecast, Makridakis (1998) proposes five steps to forecast when quantitative data is available as detailed in Fig. 4.8.

4.3.1.3 How to Improve Forecast Accuracy

Moon et al. (1998) proposed seven key points that companies should pay close attention to in order to improve its forecasting performance. The Table 4.2 summarizes each principle.

Armstrong (2001) summarizes knowledge about forecasting in 139 principles. The principles cover formulating a problem, obtaining information about it, selecting and applying methods, evaluating methods, and using forecasts. Each principle is described along with its purpose, the conditions under which it is relevant, and the strength and sources of evidence, and a checklist of principles is provided to assist in auditing the forecast process.

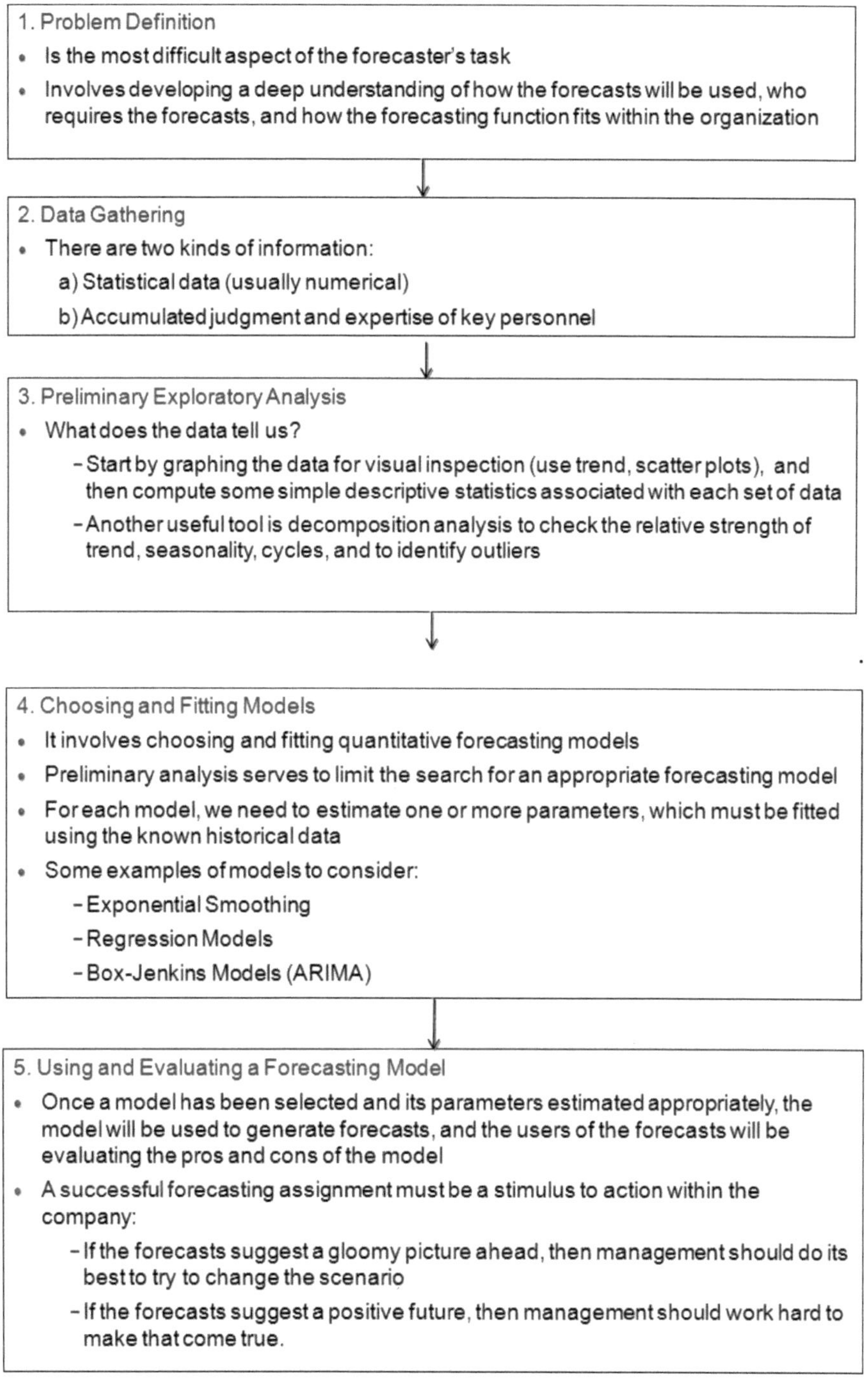

Fig. 4.8 Steps to forecast based on quantitative models (Makridakis 1998)

Table 4.2 Seven principles to better forecast (Moon et al. 1998)

Keys	Issues and symptoms	Actions	Results
Understand what forecasting is and is not	• Computer system as focus, rather than management • Blurring of the distinction between forecasts, plans, and goals	• Establish forecasting group • Implement management control system before selecting forecasting software • Derive plans from forecasts • Distinguish between forecasts and goals	• An environment in which forecasting is acknowledged as a critical business function • Accuracy emphasized and game-playing minimized
Forecast demand, plan supply	• Shipment history as the basic for forecasting demand • "Too accurate" forecasts	• Identify sources of information • Build systems to capture key demand data	• Improved capital planning and customer service
Communicate, cooperate, collaborate	• Duplication of forecasting effort • Mistrust of the "official" forecast • Little understanding of the impact throughout the firm	• Establish cross-functional approach to forecasting • Establish independent forecast group that sponsors cross-functional collaboration	• All relevant information used to generate forecasts • Forecasts trusted by users • Islands of analysis eliminated • More accurate and
Eliminate islands of analysis	• Mistrust and inadequate information leading different users to create their own forecasts	• Build a single "forecasting infrastructure" • Provide training for both users and developers of forecasts	• More accurate, relevant, and credible forecasts • Optimized investments in information/ communication systems
Use tools wisely	• Relying solely o n qualitative or quantitative methods • Cost/benefit of additional information	• Integrate quantitative and qualitative methods • Identify sources of improved accuracy and increased error • Provide instruction	• Process improvement in efficiency and effectiveness
Make it important	• No accountability for poor forecasts • Developers not understanding how forecasts are used	• Training developers to understand implications of poor forecasts • Include forecast performance in individual performance plans and reward systems	• Developers taking forecasts seriously • A striving for accuracy • More accuracy and credibility

(*continued*)

Table 4.2 (continued)

Keys	Issues and symptoms	Actions	Results
Measure, measure, measure	• Not knowing if the firm is getting better • Accuracy not measured at relevant levels of aggregation • Inability to isolate sources of forecast error	• Establish multidimensional metrics • Incorporate multilevel measures • Measures accuracy whenever and wherever forecasts are adjusted	• Forecast performance can be included in individual performance plans • Sources of errors can be isolated and targeted for improvement • Greater confidence in forecast process

These principles will also be considered as a key input to develop the maturity model.

4.3.2 Sales and Operations Planning

4.3.2.1 Introduction

Based on Wallace (2004), Sales and Operations Planning (S&OP) is a business process that helps companies keep demand and supply in balance. It does that by focusing on aggregate volumes (e.g., product families or product groups), so that, mix issues (individual products and customer orders) can be handled more readily. It occurs on a monthly cycle and displays information in both units and dollars, thus it integrates operational and financial planning.

S&OP links the company's strategic plans and business plan to its detailed processes – order entry, master scheduling, plant scheduling, and purchasing. S&OP enables the company's managers to view the business holistically, and gives them a window into the future, to avoid lack of operational capacity to meet customer demand.

S&OP is cross-functional, involving general management, sales, demand planning, manufacturing, logistics, finance, and product development. It occurs at multiple levels within the company, up to and including the executive in charge of the business unit (e.g., division president, business unit general manager, etc.).

S&OP is an integral part of the Supply Chain Management. A given supply chain will not work well if its various members do not have good volume plans in the first place, and if they are slow to react to the inevitable changes in volume. In companies without S&OP process, there is frequently a disconnection between the strategic plans and the detailed plans and schedules. In other words, the plans developed and authorized by senior management are not connected to the plans and schedules that drive day-to-day activities on the plant floor.

The framework in Fig. 4.9 clearly shows the importance of the S&OP process, as it links the strategic and business plans together with the detailed planning process at the shop floor level.

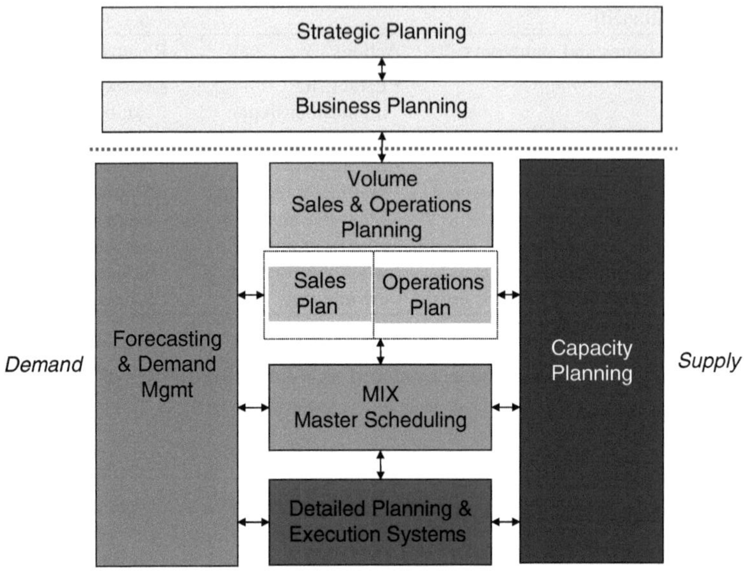

Fig. 4.9 Wallace (2004) resource planning model

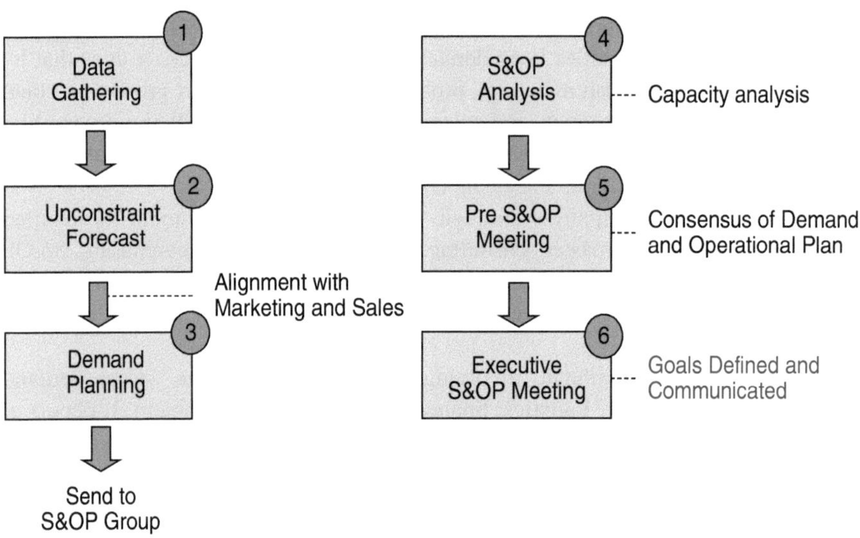

Fig. 4.10 Author's monthly Sales and Operations Planning process

4.3.2.2 Detailed S&OP Process

Sales and Operations Planning is a monthly process that comprehends six steps, as illustrated and described in Fig. 4.10.

Step 1: Data Gathering Most of this activity occurs within the Information System and/or Demand Planning departments, and happens shortly after the end of the month. It consists of three elements:

- Update the files (e.g., sales, revenue, etc.) with data from the month just ended;
- Generate information for Sales and Marketing people to use in developing the new forecast: Sales analysis data, statistical forecast reports, etc.
- Disseminate the information to the appropriate people.

Step 2: Unconstraint Statistical Forecast The second step is to generate the unconstraint forecast, and consists of two elements:

- Run statistical forecast models to predict future volumes, open by business unit, geographic regions, product family, SKU;
- Apply appropriate forecast techniques for New Products (e.g., Regression, Market Research, Sales Forecast, Conjoint Analysis)

Step 3: Demand Planning The third step is one of the most important, and refers to the alignment of the demand figures that will be used by all departments for analyzing capacity availability and operational impacts, and consists of three elements:

- Sales people review the information received in step 2, analyze, discuss and generate the forecast figures for the next period (e.g., Special for key customers like Supermarkets, where promotions have a great impact).
- New Product department review and adjust timing and volumes for new product launches.
- Document key assumptions that underlie the forecasts.

Step 4: S&OP Analysis The forth step refers to the supply (capacity) analysis and will be performed by each functional area (e.g., manufacturing, warehousing, inventory, distribution, transportation, etc.), and consists of two elements:

- Each functional area should analyze operational capacity to fulfill demand volumes:
  - Production capacity (master plan)
  - Warehousing and storage capacity
  - Inventory availability (fill rate)
  - Supplier capacity for key raw materials
  - Distribution capacity (delivery)
  - Transportation capacity (long haul)
- Estimate company's financial results based on the forecast:
  - Net Operating Profit After Tax (NOPAT), Earnings Before Interest, Taxes, Depreciation and Amortization (EBITDA), and Earnings per Share (EPS), just to give some examples.

Outputs from the S&OP analysis are the standard graphics comparing required vs. available capacity for each process, and also a list of any supply problem that cannot be resolved by the functional area or that require senior manager decision. (In some cases, demand can highly exceed supply capacity and the constraints cannot be overcome within the allowable time, requiring, for instance, extra investment.)

In some companies, they prefer to conduct a formal meeting for supply planning, while others find it more effective to simply work the process informally on a one by one basis.

Step 5: Pre S&OP Meeting The fifth step refers to the S&OP meeting, where each area will present their findings and results, and have four objectives:

- Make decisions regarding the balancing of demand and supply
- Identify areas where agreement cannot be reached, and determine how the situation will be presented in the Executive S&OP meeting
- Develop, where appropriate, alternative scenarios with different courses of action to solve a given problem
- Set the agenda for the Executive S&OP meeting

The key players in this meeting typically include several different positions like, demand manager, logistics manager, customer service, supply planning manager, production manager, and finance planning manager, sales manager and marketing manager.

Step 6: Executive S&OP Meeting This is the culminating event in the monthly S&OP cycle and has the following objectives:

- Make decisions about demand and operational plans (accept the recommendation from the Pre-S&OP team or choose another course of action)
- Authorize changes in production, procurement, distribution, where significant costs are involved
- Analyze the dollar version of the S&OP against Business Plan targets

The key players in this meeting include CEO, director of sales, marketing, supply chain, logistics, finance and human resources. Outputs from this meeting should include the meeting notes, which spell out the decisions made, and modifications to the business plan, if required.

4.3.2.3 S&OP Successful Factors

Lapide (2004) proposed a list of dozen factors that can help to implement an effective S&OP process that maintains exceptional supply chain operational performance over time. The factors are listed below and are detailed in the reference provided:

- Ongoing, routine S&OP meetings
- Structured meeting agendas
- Pre-work to support meeting inputs
- Cross-functional participation

- Participants empowered to make decisions
- An unbiased, responsible organization to run a disciplined process
- Internal collaborative process leading to consensus and accountability
- An unbiased baseline forecast to start the process
- Joint supply and demand planning to ensure balance
- Measurement of the process
- Supported by integrated supply-demand planning technology
- External inputs to the process

4.3.2.4 S&OP Maturity Model

Lapide (2005) proposed a four step maturity model to help companies assess their current performance in terms of process and enabling technology, as described in Fig. 4.11.

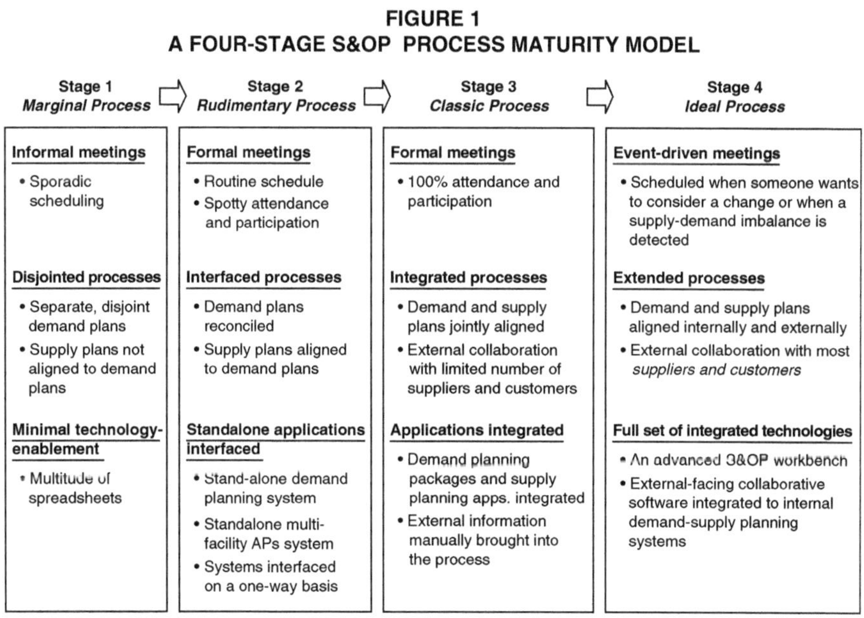

Fig. 4.11 Four step maturity model (Lapide 2005)

4.3.2.5 Benefits from S&OP Implementation

Based on Wallace (2004), the implementation of S&OP process usually results in the following benefits:

- More stable production rates and less overtime, leading to higher productivity

- Better visibility into future capacity problems, covering both over and under capacity
- Enhanced teamwork among middle-management from different areas like sales, operations, finance, etc.
- Enhanced teamwork within the executive group
- Greater accountability regarding actual performance to plan
- Better demand and supply balance across the company's supply chain
- Ability to make changes quickly off of that common game plan
- For make to stock companies: Higher customer service, and often lower finished goods inventories, at the same time
- For make to order companies: Higher customer service, and often smaller customer order backlogs and hence shorter lead times, at the same time
- For finish-to-order companies: Higher customer service, quicker response, and often lower component inventories, at the same time
- The establishment of "one set of numbers" which will be the base to run the business

Several companies, from different industries and business segments in Brazil, had already started implementation of S&OP process and concepts, as described by ILOS institute (ILOS Institute 2009) in Fig. 4.12.

Companies that implemented S&OP process achieved significant business and operational results, as listed below:

- SPP-Nemo achieved 40% increase in forecast accuracy with a 40% reduction in inventories and 27% reduction in total time to plan operations.

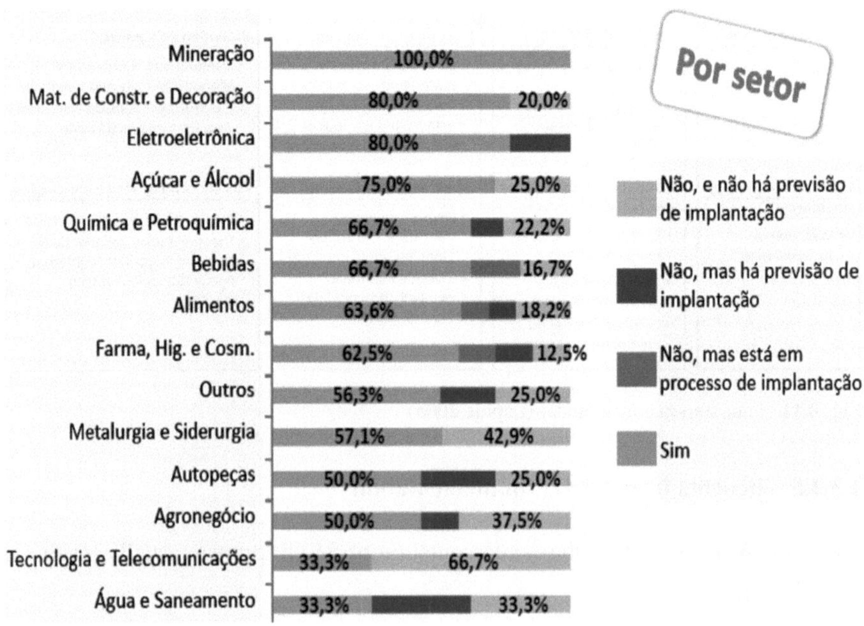

Fig. 4.12 Status of S&OP implementation in Brazil (ILOS Institute 2009)

- RECKITT BENCKISER achieved 35% reduction in forecast error based on Mean Average Percentage Error (MAPE) from 23% to 15%, and also increased visibility of customer demand, generating a unique demand plan for the whole organization.
- ARNO achieved 30% reduction in forecast error based on Mean Average Percentage Error (MAPE) from 54% to 38%, and also get greater commitment from the sales team to the planning process.
- DIMED achieved 35% reduction in forecast error based on Mean Average Percentage Error (MAPE) from 54% to 35%, and also 50% reduction in inventory of low turn products, which generated a cost avoidance of US $1 MM/year.
- MICHELIN reported an 18% reduction on average inventory levels, and also reduction of expedite imports using transportation.

4.3.3 Collaborative Planning, Forecasting and Replenishment (CPFR):

4.3.3.1 Introduction

Based on **V**oluntary **I**nterindustry **C**ommerce **S**tandards (VICS) definition, CPFR is a business practice that combines the intelligence of multiple trading partners in the planning and fulfillment of customer demand. CPFR links sales and marketing best practices, such as category management, to supply chain planning and execution processes to increase product availability while reducing inventory, transportation and logistics costs.

The CPFR model has a general framework, illustrated in Fig. 4.13, by which a buyer and seller can use collaborative planning, forecasting, and replenishing processes in order to meet customer demand. To increase performance, the buyer and seller are involved in four collaboration activities that are listed in logical order, but companies often engage in these activities simultaneously.

Strategy and Planning

The first collaboration task under this activity is Collaboration Arrangement, which is a method for defining the relationship in terms of establishing business goals, defining the scope, and assigning checkpoints and escalation procedures, roles, and responsibilities. The retailer task related to this collaboration task is *Vendor Management*, and the manufacturer task is Account *Planning*.

The second collaboration task is Joint Business Plan. This task pinpoints the major actions that affect supply and demand in the planning period. Examples of these are introducing new products, store openings and closings, changing inventory policy, and promotions. The retailer task associated with this is *Category Management* and the manufacturer task is *Market Planning*.

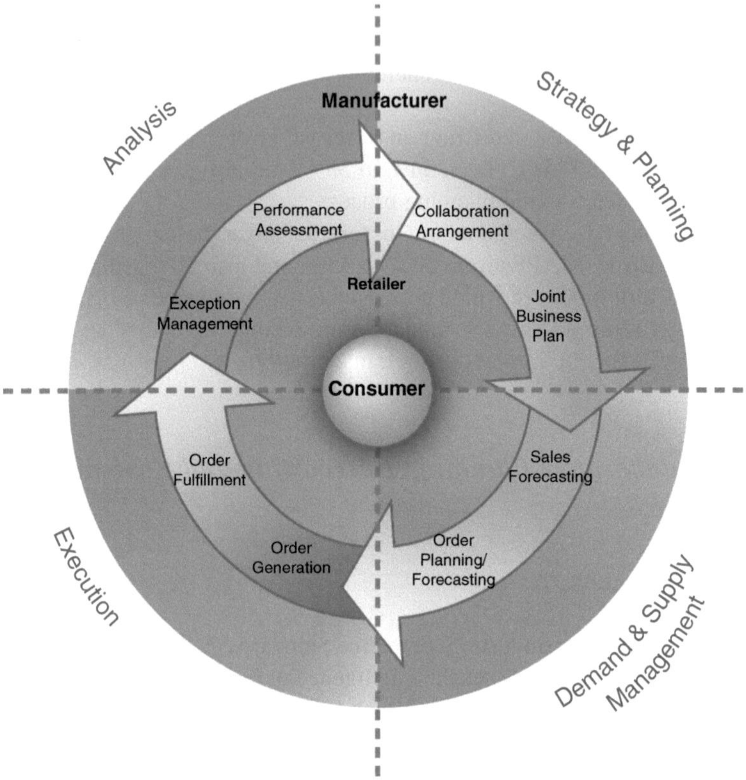

Fig. 4.13 VICS CPFR model

Demand and Supply Management

Sales forecasting, which projects point-of-sale consumer demand, is one of the collaboration tasks associated with this activity. The retailer task here is Point of Sale (*POS*) *Forecasting* and the manufacturer task is *Market Data Analysis*. The other collaboration task is Order Planning/Forecasting which uses factors, such as transit lead times, sales forecast, and inventory positions to determine future product ordering and requirements for delivery. The associated retailer task is *Replenishment Planning*, and *Demand Planning* is the associated manufacturer task.

Execution

The first collaboration task under the Execution activity is Order Generation. This task transitions forecasts to demand for the firm. The retailer task related to this collaboration task is *Buying/Replenishing*, and the manufacturer task is *Production and Supply*. The second collaboration task is Order Fulfillment, and this is the preparation of products for customer purchase through the process of producing, shipping, delivery, and stocking. In this case, both the retailer and manufacturer task is *Logistics/Distribution.*

Analysis

Exception Management, which oversees the planning and operations for conditions that are out-of-bounds, is one of the collaboration tasks associated with this activity. The retailer task is *Store Execution,* and the manufacturer task is *Execution Monitoring.*

The other collaboration task is Performance Assessment which calculates important metrics in order to discover trends, develop other strategies, and assess the attainment of business goals. The retailer task here is *Supplier Scorecard,* and the manufacturer task is *Customer Scorecard.* The model described above is a two-tiered model. However, this model can be extended to include more than two layers in the supply chain. VICS calls this "N-tier Collaboration," which is a relationship that develops from retailers through manufacturers/distributors to suppliers.

4.3.3.2 Detailed CPFR Process

With CPFR, trading partners agree to develop a collaborative business relationship based on exchanging information to support the synchronization of activities and to deliver products in response to market demand.

The following nine steps illustrated in Fig. 4.14 for CPFR implementation are based on VICS CPFR Voluntary Guidelines. Broadly, the nine steps can be further classified under three different phases – (1) Planning, (2) Forecasting and (3) Execution. The planning stage involves preparation to evaluate a company's internal requirements and capabilities, trading partner segmentation, and implementation strategy. The forecasting phase involves steps, such as creation of sales and order forecast, and exception handling, which is an ongoing iterative process. In the third phase, order execution and delivery are handled. In all three phases, trading partners work together to achieve common goals defined in the initial phase.

Phase 1: Planning

Step 1 – Develop CPFR Front-End Agreement. The entities involved in a collaborative relationship (suppliers and buyers) establish guidelines and rules for the collaborative relationship. The front-end agreement addresses each party's expectations, and the actions and resources necessary for success. To accomplish this, the two parties co-develop a general business agreement that includes the overall understanding and objective of the collaboration, confidentiality agreements, and the empowerment of resources (both actions and commitment) to be employed throughout the CPFR process.

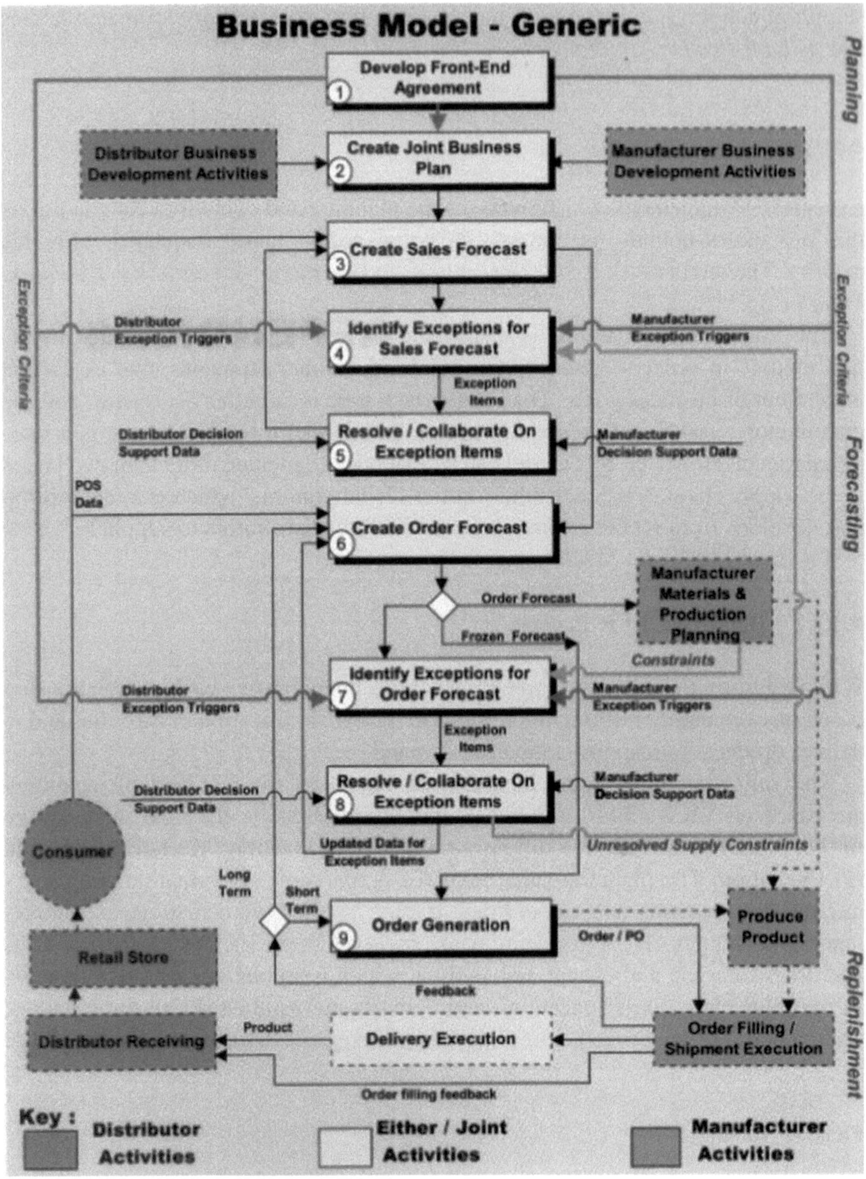

Fig. 4.14 VICS CPFR business model

Step 2 – Create Joint Business Plan. In this step of the CPFR process, the entities (suppliers and buyers) exchange information about their corporate strategies and business plans in order to collaborate on developing a joint business plan. The partners first create a partnership strategy, and then, define category roles, objectives, and tactics. The item management profiles (e.g., order minimums and multiples, lead times, order intervals) for items to be collaborated on are established.

Table 4.3 Key scenarios lead role (VICS)

Scenario	Sales forecast	Order forecast	Order feneration
A	Buyer	Buyer	Buyer
B	Buyer	Seller	Seller
C	Buyer	Buyer	Seller
D	Seller	Seller	Seller

Phase 2: Forecasting

Step 3 – Create Sales Forecast. In this step, retailer point of sale (POS) data, causal information, and information on planned events are used to create a sales forecast that supports the joint business plan. Table 4.3 describes that in scenarios A, B, and C, this step is carried out by the retailer/distributor (or buyer), and in Scenario D, the manufacturer (or seller) is responsible for creating the sales forecast.

The sales forecast is generated by either or both parties for a given period with forecasting tools that use all the relevant information and set guidelines.

Step 4 – Identify Exceptions for Sales Forecast. This step identifies the items that fall outside the sales forecast constraints set jointly by the manufacturer and distributor. (The exception criteria for each item are agreed to in the Front-end agreement).

Step 5 – Resolve/Collaborate on Exception Items. This step involves resolving sales forecast exceptions by querying shared data, email, telephone conversations, meetings, and so on, and submitting any resulting changes to the sales forecast. Collaborative negotiations between buyer and sellers resolve item exceptions.

Step 6 – Create Order Forecast. In this step, POS data, causal information, and inventory strategies are combined to generate a specific order forecast that supports the shared sales forecast and the joint business plan. Actual volume numbers are time-phased and reflect inventory objectives by product and receiving location. The short-term portion of the forecast is used for order generation, while the longer-term portion is used for planning.

Step 7 – Identify Exceptions for Order Forecast. This step determines what items fall outside the order forecast constraints set jointly by the manufacturer and distributor.

Step 8 – Resolve/Collaborate on Exception Items. This step involves the process of investigating order forecast exceptions through querying of shared data, email, telephone conversations, meetings, and so on, and submitting any resulting changes to the order forecast.

Phase 3: Execution

Step 9 – Order Generation. This last step marks the transformation of the order forecast into a committed order. Order generation can be handled by either the manufacturer or distributor depending on competencies, systems, and resources.

Regardless of who completes this task, the created order is expected to consume the forecast.

4.3.3.3 CPFR Technology

The CPFR process does not fundamentally depend upon technology. However, specialized technology can make the process more scalable. Many CPFR solutions have been developed to facilitate the process, including:

- Sharing forecasts and historical data
- Automating the collaboration arrangement and joint business plan
- Evaluating exception conditions
- Enabling revisions and commentary

A CPFR solution must be integrated with the enterprise systems of record that produce and consume demand and supply chain data, as illustrated in the Fig. 4.15.

CPFR technology can be deployed as a shared solution, or as a peer-to-peer network of interoperating CPFR applications. The shared solution can be operated as part of a retailer's or manufacturer's extranet, or hosted by an exchange or other third party. Peer-to-peer communications may flow directly between manufacturers and suppliers, or via proxies (trading-partner-to-exchange or exchange-to-exchange).

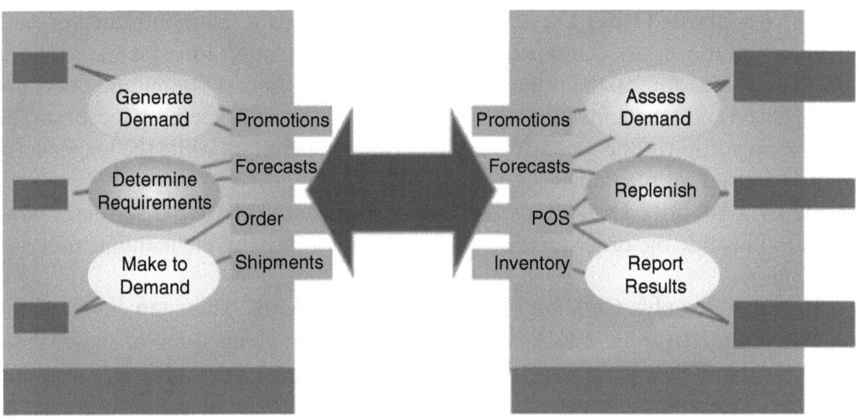

Fig. 4.15 The role of CPFR technology in integrating retailer and manufacturer processes

4.3.3.4 Benefits

Based on Oliver Wight (2005) research, the implementation of CPFR process usually results in the following benefits:

- Forecast accuracy improvements of 20–30%
- Sales revenue growth of 8–10%
- Cost of goods sold reduced 3–4%
- Operating costs reduced 1–2%
- Lead times and cycle times reduced 25–30%

4.3.3.5 CPFR Assessment

The Voluntary Interindustry Commerce Standards (VICS) developed a CPFR capability assessment to provide a framework for understanding the gap between a company's existing practices and CPFR best practices, and also to serve as the starting point for change and enable realistic expectations for a CPFR program.

Table 4.4 shows the four key areas covered in the VICS assessment and illustrates the increasing difficulty and benefits of progressing vertically through the processes of **Collaboration** to **Integrated Planning and Forecasting** to **Replenishment** and finally to **Supply Chain Management.**

It will not be presented the detailed VICS assessment in the body of this book. However, it will be considered as a key input when developing the Demand Driven Supply Chain Maturity Model in Chap. 5.

4.3.4 Vendor Managed Inventory and Demand Visibility

4.3.4.1 Introduction

VMI is essentially a distribution channel operating system whereby the inventory at the distributor/retailer is monitored and managed by the manufacturer/vendor. It includes several tactical activities including, determining appropriate order

Table 4.4 VICS capability assessment

Process area	Basic	Developing	Advanced
D Supply chain management	No supply chain focus/plan	Internal enterprise optimization	Supply chain optimization
C Replenishment processes	Pre-DC limited/no retail visibility	DC replenishment focus	Computer assisted retail ordering flow-through
B Integrated planning and forecasting processes	Manual non-standard forecasting planning	Standardized demand data creation and input	
A Collaborative processes	Limited one-way communication	Standardized and integrated collaboration	Integrated planning. Forecasting and collaboration

quantities, managing proper product mixes, and configuring appropriate safety stock levels. The rationale is that by pushing the decision making responsibility further up the supply chain, the manufacturer/vendor will be in a better position to support the objectives of the entire integrated supply chain, resulting in a sustainable competitive advantage. Centralizing the replenishment decision also helps reduce the distortions in ordering introduced when there are several intermediaries that place orders in a supply chain.

VMI was popularized in the late 1980s by Wal-Mart and Procter & Gamble (Waller et al. 1999), and then, it was subsequently implemented by many other leading companies from different industries, such as GlaxoSmithKline (Danese 2004), Electrolux Italia (De Toni and Zamolo 2005), Nestle and Tesco (Watson 2005), etc.

The enabling technology behind successful VMI is Electronic Data Interchange (EDI) which provides manufacturer/vendor with essentially the same point of sales (POS) and inventory information retained by the distributor/retailer. As a result, improved forecasting is possible because the manufacturer/vendor can observe demand for its product over a wider range of customers and can incorporate the effects of promotions, competing products, and seasonal variations in demand. Therefore, successfully integrating systems technology in the transactions between value chain participants is integral to realize benefits of VMI.

4.3.4.2 Detailed VMI and Demand Visibility Process

Based on Hall (2001), there are two EDI (Electronic Data Interchange) transactions at the heart of the process as illustrated in Fig. 4.16.

The first is the "Product Activity Record," frequently referred to as an 852. The data contained in this document are sales and inventory information. The inventory data is typically segmented into various groups, such as on hand, on order, committed, back ordered, and so forth. This transaction is the backbone of VMI, and is sent by the customer on a prearranged schedule, typically, daily. The decision to order is based on this data.

The business process fed by this data is relatively simple. The supplier reviews the information that has been sent in by the customer on the 852 to determine if an order is needed. This review of the data varies by supplier and the software being used, but, many things are consistent:

- The first step is to verify if the data is accurate and meaningful. Depending on the software, much of this verification is automated.
- On a scheduled basis, the software calculates a reorder point for each item based on the movement data and any overrides contributed by the customer or supplier. These overrides might include information such as promotions, projects, seasonality, new items, etc.
- The VMI software compares the quantity available at the customer with the reorder point for each item at each location (SKU by location). This determines if an order is needed.

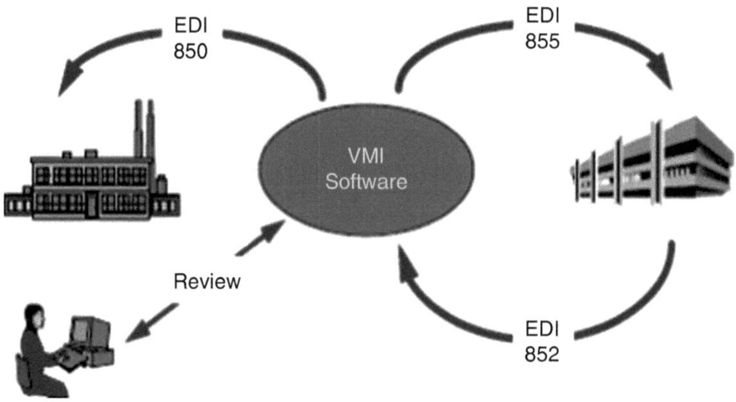

Fig. 4.16 EDI documents for VMI Implementation

- The order quantities are then calculated, and typically calculation of order quantities takes into account such issues as case quantities and transaction costs. This completes the order build process.

The second VMI transaction informs the customer what product will be delivered by the supplier. There are two transactions being used for this function. The most frequently used is the "purchase order acknowledgment," referred to as the 855. This document contains the product numbers and quantities ordered by the supplier on the customer's behalf. A few customers skip the 855 and rely on the advance ship notice (ASN), or 856, to alert them to the order and shipment. This document differs from the 855 in both timing and content. The 856 is sent after the shipment has been made instead of at the time of the order. The 856 contains only the part numbers shipped as well as additional information, such as carrier and waybill information. For the purposes of VMI, either of these documents works well if properly implemented.

4.3.4.3 Benefits

Based on the working paper from Kellogg (Kellogg Graduate School of Management 2000), VMI can have a number of benefits, including lowered investment in the supply chain, due to better forecasting, JIT delivery and less overstocking and greater inventory turnover. Its primary benefit, however, is improved customer service due to fewer stock outs and more optimal product mixes. Manufacturer/ vendor also stand to benefit from VMI, as it allows them to schedule production and transportation more efficiently (including ordering raw materials), to observe end-user consumption and general market trends more closely, and to develop

Table 4.5 Benefits from VMI implementation (Kellogg University 2000)

Typical benefits to Mfg/Vend	Typical benefits to Dist/Ret
• Lower inventory investment (raw and finished)	• Fewer stock-outs with higher turnover
• Better scheduling and planning	• Better market information
• Better market information	• More optimal product mixes
• Closer customer ties and preferred status	• Less inventory in channel (transfer costs)
	• Lower administrative replenishment costs

closer ties with their customers. In summary, the benefits of VMI Program are detailed in Table 4.5.

The working paper from Kellogg also reports the following benefits out of the VMI implementation:

- Fred Meyer, the 131-unit chain of supercenters in the Pacific Northwest, reduced inventories 30–40%, while sales rose and service levels increased to 98%. This was due to a VMI program implemented with two key food vendors.
- Grand Union, a New Jersey-based grocery retailer with more than 100 stores and three DCs, improved inventory turns by close to 80% and achieved 99% service levels. This significantly improved sales by eliminating out-of-stock conditions and dramatically reduced warehousing costs.
- Oshawa Foods, a $6 billion Canadian food distributor and retailer, had tremendous success with Pillsbury, Quaker and H.J. Heinz with inventory turns improving from 3 to 9 times, while achieving customer service levels of 99%.

The author had also implemented a VMI project in the beverage industry with one key customer from the airline industry to increase customer service level (fill rate) at the same time that optimizes the inventory levels. Figure 4.17 illustrates the relationship between the two companies:

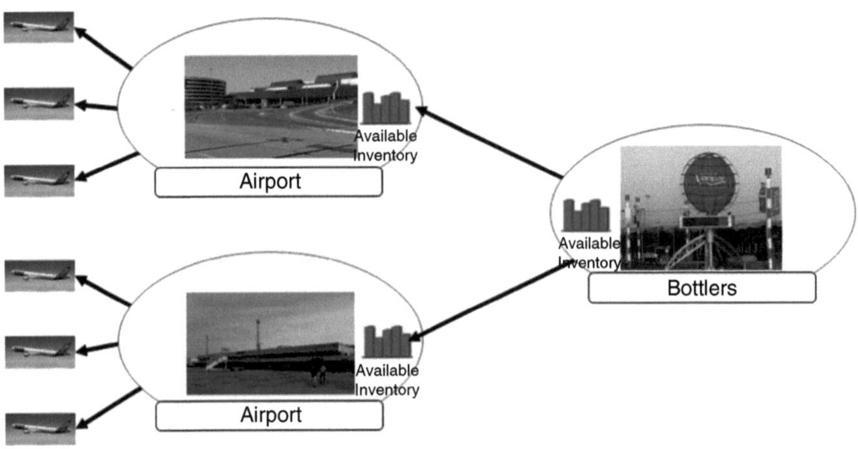

Fig. 4.17 VMI project implemented in the beverage industry

The project started in Feb/2008 and 4 months later (June/2008), the project team was able to measure 55% reduction in inventory level with 100% fill rate in this period, as illustrated in the Fig. 4.18.

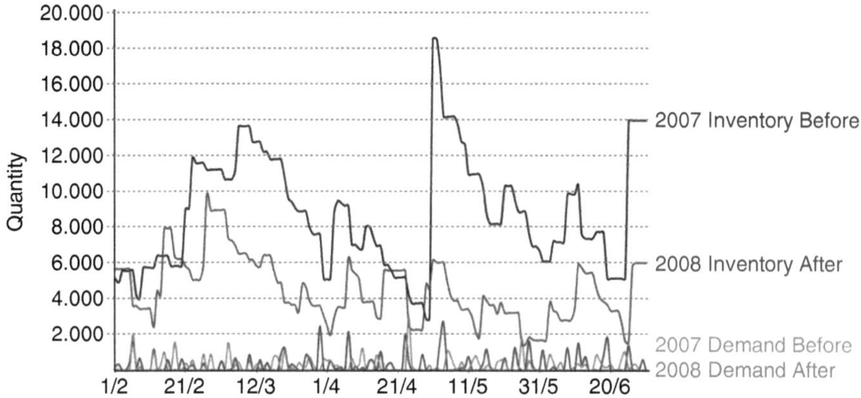

Fig. 4.18 Daily demand and inventory levels before and after VMI project

4.4 Supply and Operations Management

In this section, it will be performed a literature review for each one of the five categories of the Supply & Operations management – Procurement, Manufacturing, Logistics, Customer Service and Senior Management Support. This review allowed identify the DDSC characteristics for each category which was used to develop the five level maturity model.

4.4.1 Introduction

Supply and Operations Management refers to the capabilities of the firm to source, produce, store, sell and delivery its products in the market place. It is a critical capability both in terms of cost, due to all fixed and variable costs required to perform the operational activities, and also in terms of customer service, due to the high pressure of customers towards better and customized services.

Stewart (1997), states that managing supply-chain operations is critical to any company's ability to compete effectively, and that success for many companies now depends on their ability to balance a stream of product and process changes with meeting customer demands for delivery and flexibility. Optimally managing supply-chain operations, has therefore, become critical to companies' ability to compete effectively in the global marketplace.

Still based on (Stewart 1997), to assist companies in increasing the effectiveness of their supply chain, and to support the move to process-based management, two consulting firms – PRTM and Advanced Manufacturing Research (AMR) – set out to consolidate within a process reference model their experience along with a group of senior operations, manufacturing and supply chain managers from many of the leading companies. This group of companies, together with other leading US and multinational firms, joined together in 1996 to form the Supply-Chain Council (SCC). The SCC took the reference model and helped develop, test and finally release it, calling it the Supply Chain Operations Reference Model (SCOR). SCOR is the first cross-industry framework for evaluating and improving enterprise-wide supply-chain performance and management.

SCOR is designed to enable companies to communicate, compare and develop new or improved supply-chain practices from companies both within and outside of their industry segment. Its key components are:

- Standard descriptions of the process elements that make up complex management processes
- Benchmark metrics used to compare process performance to objective, external points of reference
- Description of best-in-class management practices
- Mapping of software products that enable best practices

SCOR model spans:

- All customer interactions, from order entry through paid invoice
- All physical material transactions, from the supplier's supplier to the customer's customer, including field service logistics
- All market interactions, from the understanding of aggregate demand to the fulfillment of each order

SCOR model focuses on five basic supply chain processes, as described below:

- **Plan – Demand/Supply Planning and Management**
  - Balance resources with requirements and establish/communicate plans for the whole supply chain, including return, and the execution processes of Source, Make and Deliver
  - Management of business rules, supply chain performance, data collection, inventory, capital assets, transportation, planning configuration, regulatory requirements and compliance, and supply chain risk
  - Align the supply chain unit plan with the financial plan
- **Source – Sourcing Stocked, Make to Order, and Engineer-to-Order**
  - Schedule deliveries, receive, verify and transfer products, and authorize supplier payments
  - Identify and select supply sources when not predetermined, as for engineer-to-order product
  - Manage business rules, assess suppliers' performance, and maintain data
  - Manage inventory, capital assets, incoming product, supplier network, import/export requirements, supplier agreements, and supply chain source risk

- **Make – Make to Stock, Make to Order, and Engineer to Order Production Execution**
  - Schedule production activities, issue product, produce and test, package, stage product, and release product to deliver
  - Finalize engineering for engineer-to-order product
  - Manage rules, performance, data, work-in-process products (WIP), equipment and facilities, transportation, production network, regulatory compliance for production, and supply chain make risk
- **Deliver – Order, Warehouse, Transportation/Distribution**
  - All order management steps from processing customer inquiries and quotes to routing shipments and selecting carriers
  - Warehouse management from receiving and picking product to load and ship product
  - Receive and verify product at customer site, and install, if necessary
  - Invoicing customer
  - Manage deliver business rules, performance, information, finished product inventories, capital assets, transportation, product lifecycle, import/export requirements, and supply chain deliver risk
- **Return – Return of Raw Materials and Receipt of Returns of Finished Goods**
  - All return defective product steps from source – identify product condition, disposition product, request product return authorization, schedule product shipment, and return defective product – and deliver – authorized product return, schedule return receipt, receive product, and transfer defective product
  - All return excess product steps from source – identify product condition, disposition product, request product return authorization, schedule product shipment, and return excess product – and deliver – authorize product return, schedule return receipt, receive product, and transfer excess product
  - Manage return business rules, performance, data collection, return inventory, capital assets, transportation, network configuration, regulatory requirements and compliance, and supply chain return risk

For the sake of supply and operations management in this book, the supply chain processes proposed by SCOR will be covered, but using a different terminology, as detailed below, in order to be closer to the organization structure names found in most industries.

4.4.2 Procurement

Lockamy and McCormack (2004) performed an exploratory study to investigate the link of SCOR planning processes to supply chain performance. In this study, they showed that supplier transactional collaboration activities have a significant impact on supply chain performance within the SOURCE decision area. These activities include the sharing of planning and scheduling information with suppliers. The source planning process, which includes the documentation of procurement processes, the establishment of information technology that supports these processes, and the management of supplier inter-relationships, also has a significant impact on supply chain performance in this decision area.

Supplier inter-relationships included in the source planning process include the management of product and delivery variability, along with metrics for monitoring such variability. Additionally, the designation of a source planning process owner is required to ensure its effectiveness. The establishment of a procurement process planning team was found to have an impact on supply chain performance within the SOURCE decision area. This team should meet on a regular basis, and work closely with other functional areas, such as manufacturing and sales. Supplier operational collaboration also has a significant impact on supply chain performance. This involves the development of a joint operational plan that is supportive of strategic sourcing activities, and outlines how routine transactional activities are to be conducted by the participants.

Supplier strategic collaboration activities also impact supply chain performance in the Source decision area. These activities include electronic ordering and supplier-managed inventory. In addition, the presence of on-site employees of key suppliers facilitates strategic supplier collaboration activities that enhance overall supply chain performance.

Ayers (2006) provides a seven steps methodology to allow companies embark on the journey from purchasing to strategic procurement. The steps are:

- **Step 1: Determine your spending**. In this step, companies should quantify how much was spent, who spent the money, where and how was it spent and what specifically was it spent for.
- **Step 2: Prioritize the spend categories**. Prioritizing means looking at the size of the savings opportunities compared to the degree of difficulty in actually achieving the savings. The degree of difficulty is determined by such issues as organizational turf, complexity of the product or service, and the complexity of the actual sourcing process (e.g., vendor selection, vendor certification, Request for Proposal – RFP, negotiations, etc.). He suggested to always starting with the "low-hanging fruit," which are the opportunities with high savings and that can be achieved with little difficulty.
- **Step 3: Form category teams**. These teams are small groups with the charter to examine the sourcing options for the category and to make recommendations to senior management. Ideally, especially in a decentralized organization, teams should have representatives from each of the key business units to obtain diversified inputs and to build consensus.
- **Step 4: Develop a sourcing strategy for your categories**. Each category team needs to develop a basic strategy to source its category. These can range from joint ventures with suppliers when the product or service being sourced is highly

technical, critical to your business, and only a few suppliers are capable of meeting the specifications to very competitive bidding situations, when the product is simple and widely available.

- **Step 5: Perform the Request for Proposal (RFP) process and make the final selection**. Identify a list of potential suppliers, starting with your current suppliers. The category team should reach consensus on the basic ground rules for awarding the business (e.g., national contract for all corporation's office supply) and the criteria (e.g., prices, rebates, breadth of offering, delivery frequencies, etc.) to use to select the wining proposal. It is extremely important to reach consensus on these issues before sending out the RFP (Request for Proposal).
- **Step 6: Manage the supplier relationship aggressively**. Supplier management is the area of strategic sourcing with the greatest opportunity for both success and failure. Too many companies just sign the contract and forget about the relationship until contract renewal time. To make the relationship a real success, ensure that the benefits you and the category team fought so hard to achieve are sustained. Both parties should be actively involved in monitoring results, reviewing pre-established performance metrics, partnering on creative ways to mutually lower costs, and ironing out any contract or performance disputes.
- **Step 7: Provide feedback to both suppliers and senior management**. Category team should provide regular feedback to suppliers on both successes and failures. It is important to make them feel a part of company's overall strategic sourcing process and also keep senior management informed about what the team has accomplished. Present an annual plan to senior management that recaps savings achieved during the year and planned activities for the upcoming year.

Harrison (2003) states that there are two extremely different approaches for managing the relationship with suppliers: On line procurement (also called "eProcurement") and Strategic Alliances. He claims that the decision on which approach to take should be based on the characteristics of the purchased component and of the marketplace. A summary of the risks and benefits of each of these two approaches is provided in Tables 4.6 and 4.7.

4.4.3 Manufacturing

Ayers (2006) provides a roadmap to implement demand driven supply chain concepts in the manufacturing area, which is illustrated in Fig. 4.19.

Evolution to a demand-driven supply chain will likely proceed in the order these items are listed. Shortening the lead-time is fundamental to changing batch model economics. Basing decisions on demand comes after adopting the economics of the flow model. Along the path there is feedback to earlier steps.

Table 4.6 Risks and benefits of strategic alliances (Harrison 2003)

Benefits to buyer	Risks to buyer	Benefits to supplier	Risks to supplier
Decrease total cost of ownership	Increased transactions cost per supplier	Locks in the business	Limited opportunities for new business, particularly with alliance partner's competitors
Increased quality	Supplier becomes monopolistic, less responsive	Ability to increase skill	Capacity locked up by partner
Faster response		Ability to make long-term investments	
Enhanced new product development with supplier involvement		Higher margins	
Highly skilled supplier base			
Fewer suppliers to manage			

Table 4.7 Risks and benefits of on line procurement (Harrison 2003)

Benefits to buyer	Risks to buyer	Benefits to supplier	Risks to supplier
Decreased unit cost	Decreased quality	Access to new business	Lower margins
Decreased transactions and processing cost	Loose specifications	Use of excess capacity	Decreased ability to invest in improvements
Faster response	Fewer suppliers over the long term	Knowledge of winning bid	Startup costs for new software
	Alienate suppliers		Buyer uses information to generate off-line bids

4.4.3.1 Cells

Cells are a proven way to save time and reduce cost in both manufacturing and service companies. The design of any production process will fall into one of two generic categories:

- Functional design – It is the traditional approach and it has its foundation in traditional accounting mentality and the associated batch approach to production. In a functional setup, the work must progress sequentially through each unit. Because the batch model stresses worker and machine utilization while ignoring lead-time, the functional design is a natural response despite not so obvious penalties in terms of lead-time.
- Cell design – Cell design is a component of the Toyota Production System, or lean manufacturing. In the cellular shop floor, machines of different types are located together and the focus shifts to the product, not the means of production. In a cellular paperwork process, small groups do all operations.

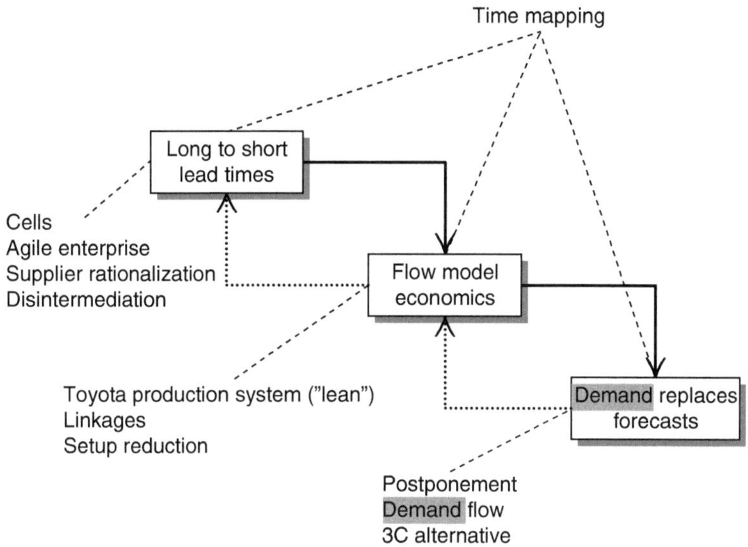

Fig. 4.19 Three phase roadmap for implementing DDSC in manufacturing (Ayers 2006)

Cells reduce lead-times and enable products to be produced in small-lot or single-lot quantities. There are three outcomes from implementing cells in manufacturing environment: improvement in flow, density and velocity, as described below:

- Flow should include simplicity of layout and minimal movement distance of components around the factory.
- Density is the measure of workspace to total space, that is, how much of the factory floor is occupied by value-adding activities as a proportion of total space. Density should be as high as possible.
- Velocity refers to the percent of time spent in value-added operations. Low velocities are characteristic of high waits for processing – usually in a batch process.

Cellular manufacturing has other benefits like:

- Improved quality – In a functional batch manufacturing setup, a whole batch of bad product may be produced before the error is detected. In the high-velocity cellular environment, the next operation will receive the product much more quickly, and defects will be caught before more bad products are produced. Feedback to the operator producing the bad parts is also fast, facilitating a learning environment.
- Focused factory – Cells also facilitate implementation of the "focused factory," a concept developed by Wickham Skinner, and which characterizes a factory that does not attempt to do much different things. The analogy is the athlete. To excel in a sport, one must concentrate on that sport. The multisport professional athlete is relatively rare. Within the focused factory, each cell can be tailored to customer needs, rather than operating in a "one-size-fits-all" environment.

4.4.3.2 Agile Enterprise

The agile enterprise line of thinking relates to the need to respond quickly in dynamic markets. In these markets, opportunities come and go rapidly, and to respond, managers must design production systems capable of rapid deployment to meet these opportunities.

The agile enterprise will be able to do two things well:

- Rapidly reposition internal operations for new opportunities, which mean structuring internal operations using cells and focused factories.
- Be good at partnerships, which mean be ready to form or participate in multi-company supply chains.

Agility requires rapid responses. Cell capabilities could be deployed more rapidly than entire factories, and so, the focus in cell design may be more on building in the ability to shift market positions rather than defend any particular market position. Flexibility, rather than cost, should be the primary goal.

The agile enterprise extends the philosophy of cellular manufacturing up into the organization. It would encompass infrastructure, including control systems, union contracts, rewards and incentives, information systems, and technology competencies. This means, in all probability, that the focused factory or cell can even be dissociated from the enterprise. The agile enterprise will be fast, not only in putting out product once it is up and running, but also in setting up supply chains in response to market opportunities.

4.4.3.3 Toyota Production System (Lean Manufacturing)

The Toyota Production system (TPS) is a philosophy of manufacturing that Toyota credits with its success in producing high quality automobiles. The concepts behind TPS underpin the "lean manufacturing model."

We can define Lean as "*. . . A systematic approach to identifying and eliminating waste through continuous improvement, flowing the product at the pull of the customer in pursuit of perfection. . .*".

The wastes noted above are commonly referred to as "non-valued-added" activities, and are known to Lean practitioners as the "Eight Wastes." Taiichi Ohno (co-developer of the Toyota Production System) suggests that these account for up to 95% of all costs in non-Lean manufacturing environments. These wastes are:

- Overproduction – Producing more than the customer demands. The corresponding Lean principle is to manufacture based upon a pull system, or producing products just as customers order them. Anything produced beyond this (buffer or safety stocks, work-in-process inventories, etc.) ties up valuable labor and material resources that might otherwise be used to respond to customer demand.

- Waiting – This includes waiting for material, information, equipment, tools, etc. Lean demands that all resources are provided on a just-in-time (JIT) basis – not too soon, not too late.
- Excess of Transportation – Material should be delivered to its point of use. Instead of raw materials being shipped from the vendor to a receiving location, processed, moved into a warehouse, and then transported to the assembly line, Lean demands that the material should be shipped directly from the vendor to the location in the assembly line where it will be used. The Lean term for this technique is called "point-of-use-storage" (POUS).
- Non-Value-Added-Processing – Some of the more common examples of this are reworking (the product or service should have been done correctly the first time), deburring (parts should have been produced without burrs, with properly designed and maintained tooling), and inspecting (parts should have been produced using statistical process control techniques to eliminate or minimize the amount of inspection required). A technique called Value Stream Mapping is frequently used to help identify non-valued-added steps in the process (for both manufacturers and service organizations).
- Excess of Inventory – Related to Overproduction, inventory beyond that needed to meet customer demands negatively impacts cash flow and uses valuable floor space. One of the most important benefits for implementing Lean Principles in manufacturing organizations is the elimination or postponement of plans for expansion of warehouse space.
- Defects – Production defects and service errors waste resources in four ways. First, materials are consumed. Second, the labor used to produce the part (or provide the service) the first time cannot be recovered. Third, labor is required to rework the product (or redo the service). Fourth, labor is required to address any forthcoming customer complaints.
- Excess of Motion – Unnecessary motion is caused by poor workflow, poor layout, housekeeping, and inconsistent or undocumented work methods. Value Stream mapping (see above) is also used to identify this type of waste.
- Underutilized People – This includes underutilization of mental, creative, and physical skills and abilities, where non-Lean environments only recognize underutilization of physical attributes. Some of the more common causes for this waste include – poor workflow, organizational culture, inadequate hiring practices, poor or non-existent training, and high employee turnover.

In order to reduce or eliminate the above wastes, Lean practitioners utilize many tools or Lean Building Blocks. Successful practitioners recognize that, although most of these may be implemented as stand-alone programs, few have significant impact when used alone. Additionally, the sequence of implementation affects the overall impact, and implementing some out of order may actually produce negative results (for example, you should address quick changeover and quality before reducing batch sizes). The more common building blocks are listed below:

- **Pull System** – The technique for producing parts at customer demand. Service organizations operate this way by their very nature. Manufacturers, on the other hand, have historically operated by a Push System, building products to stock (per sales forecast), without firm customer orders.

- **Kanban** – A method for maintaining an orderly flow of material. Kanban cards are used to indicate material order points, how much material is needed, from where the material is ordered, and to where it should be delivered.
- **Work Cells** – The technique of arranging operations and/or people in a cell (U-shaped, etc.) rather than in a traditional straight assembly line. Among other things, the cellular concept allows for better utilization of people and improves communication.
- **Total Productive Maintenance** – TPM capitalizes on proactive and progressive maintenance methodologies and calls upon the knowledge and cooperation of operators, equipment vendors, engineering, and support personnel to optimize machine performance. Results of this optimized performance include elimination of breakdowns, reduction of unscheduled and scheduled downtime, improved utilization, higher throughput, and better product quality. Bottom-line results include lower operating costs, longer equipment life, and lower overall maintenance costs.
- **Total Quality Management** – It is a management system used to continuously improve all areas of a company's operation. TQM is applicable to every operation in the company and recognizes the strength of employee involvement.
- **Quick Changeover (Setup Reduction and Single Minute Exchange of Dies – SMED)** – The technique of reducing the amount of time to change a process from running one specific type of product to another. The purpose for reducing changeover time is not only for increasing production capacity, but also to allow more frequent changeovers in order to increase production flexibility. Quicker changeovers allow for smaller batch sizes.
- **Batch Size Reduction** – Historically, manufacturing companies have operated with large batch sizes in order to maximize equipment utilization, assuming that changeover times were "fixed" and could not be reduced. Because Lean calls for the production of parts to customer demand, the ideal batch size is ONE. However, a batch size of one is not always practical, so the goal is to practice continuous improvement to reduce the batch size as low as possible. Reducing batch sizes reduces the amount of work-in-process (WIP) inventory. Not only does this reduce inventory-carrying costs, but also production lead-time or cycle time is approximately directly proportional to the amount of WIP. Therefore, smaller batch sizes shorten the overall production cycle, enabling companies to deliver more quickly and to invoice sooner (for improved cash flow). Shorter production cycles increases inventory turns and allows the company to operate profitably at lower margins, which enables price reductions, which increases sales and market share.
- **5S or Workplace Organization** – This tool is a systematic method for organizing and standardizing the workplace. It's one of the simplest Lean tools to implement, provides immediate return on investment, crosses all industry boundaries, and is applicable to every function with an organization. Because of

these attributes, it's usually the first step recommended for a company start its journey to Lean manufacturing.

As the name indicates, 5S consists of five phases, as described below:

- **Sorting** – Eliminate all unnecessary tools, parts, instructions. Team should go through all tools, materials, etc., in the plant and work area and keep only essential items. Everything else should be stored or discarded.
- **Straightening** – There should be a place for everything, and everything should be in its place. The place for each item should be clearly labeled or demarcated. Items should be arranged in a manner that promotes efficient work flow. Workers should not have to repetitively bend to access materials. Each tool, part, supply, piece of equipment, etc. should be kept close to where it will be used (i.e., straighten the flow path).
- **Systematic cleaning** – Workplace should be kept tidy and organized. At the end of each shift, workers should clean the working area and be sure everything is restored to its place. This makes it easy to know what goes where and ensures that everything is where it belongs. A key point is that maintaining cleanliness should be part of the daily work – not an occasional activity initiated when things get too messy.
- **Standardizing** – Work practices should be consistent and standardized. Everyone should know exactly what his or her responsibilities are for adhering to the first 3 S's.
- **Sustaining** – Maintain and review standards. Once the previous 4 S's have been established, they become the new way to operate. Employees should maintain focus on this new way and do not allow a gradual decline back to the old ways. While thinking about the new way, they should also be thinking about better ways to operate. When an issue arises, such as a suggested improvement, a new way of working, a new tool or a new output requirement, review the first 4 S's and make changes as appropriate.
- **Visual controls** – These are simple signals that provide an immediate and readily apparent understanding of a condition or situation. Visual controls enable someone to walk into the workplace and know within a short period of time what's happening with regards to production schedule, backlog, workflow, inventory levels, resource utilization, and quality. These controls should be efficient, self-regulating, and worker managed, and include visual boards, Kanban cards, lights, color-coded tools, lines delineating work areas and product flow, etc.

4.4.3.4 Postponement

Postponement refers to efforts to customize products for delivery as late in the production process as possible to fulfill the demand cost effectively. It supports the just in time principle from the Toyota Production System.

Designing for postponements makes the concept easier to implement. These designs create needed commonality among end products. Assembling common parts or modules creates each unique product configuration. However, implementation requires multidepartment collaboration, including operations, marketing, procurement, and engineering.

Without commonality, plants may be efficient as far as individual products are concerned. However, total cost, including inventory and shipping, will be high due to proliferation of specialized parts and finished goods inventory.

Nahmias (2005) provides examples of postponement concept implementation in Benetton Group, who operates in the fashion clothes industry and also in Hewlett-Packard (HP), who operates in the printer industry.

4.4.4 Logistics

4.4.4.1 Introduction

Bowersox and Lahowchich (2008) state that Logistics is a critical area in the responsive supply chain business model. They argue that the logistical structure include the traditional functional areas of order management, transportation, inventory, warehousing and material handling. These five dimensions of logistical performance are integrated in an operating system and a facility network design. Included within these primary areas are important strategic and tactical matters, such as protective packaging, building and material handling design, facility location and network design, as well as reverse movement related to product warranty, recall, and disposal (commonly called reverse logistics).

4.4.4.2 Warehouse

Based on Baker and Canessa (2009), warehouses are a key aspect of modern supply chains and play a vital role in the success, or failure, of businesses today.

Two types of warehouse can be distinguished: distribution warehouses and production warehouses.

- The function of a distribution warehouse is to store products and to fulfill external customer orders typically composed of a large number of order lines (where each order line specifies a quantity of one particular product). The number of different products in a distribution warehouse may be large, while the quantities per order line may be small, which often results in a complex and relatively costly order picking process. Therefore, distribution warehouse are often optimized for cost-efficient order picking. The prominent design criterion is the maximum throughput to be reached at minimum investment and operational costs.
- The function of a production warehouse is to store raw materials, work-in-process and finished products, associated with a manufacturing and/or assembly process. Raw materials and finished products may be stored for long periods.

This occurs, for example, when the procurement batch of incoming parts is much larger than the production batch, or when the production batch exceeds the customer order quantity of finished products. Storage of goods for long periods must be cost-efficient, and is usually done in large quantities in an inexpensive storage system, such as pallet rack. The prominent design objectives are low investment costs and operational costs.

Huertas et al. (2007) state that facility layout plays an important role in the business success of the company, and the most appropriate warehouse layout depends on its particular operational conditions, and characteristics such as modularity, adaptability, compactness, distribution of movements, accessibility and flexibility. Layout design is a problem-dependant, in other words, there is no best design, methodology or policy for all problems under consideration. Selecting the best layout for a given case is not trivial, because of the diversity of factors influencing a warehouse operation, such as docks location, aisles access, racks types, racks access.

Baker and Canessa (2009) also provide an overall framework with the key steps and main tools used, as well as the literature references on warehouse design, as illustrated in Table 4.8.

Gu et al. (2010) state that warehouse design involves five major decisions:

- Determine the overall structure: The overall structure (or conceptual design) determines the functional departments, such as how many storage departments, technology employed and how orders will be assembled. At this stage of the design, the issues are to meet storage and throughput requirements, and to minimize costs.
- Department layout: It is the detailed configuration within a warehouse department, for example, aisle configuration in the retrieval area, pallet block-stacking pattern in the reserve storage area, and configuration of an automated storage/retrieval systems (AS/RS).
- Operation strategy: Refers to those decisions about operations that have global effects on other design decisions, and therefore, need to be considered in the design phase. Examples are the decision between randomized and dedicated storage, or the decision to use zone picking.
- Equipment selection: Determine an appropriate automation level for the warehouse, and identify equipment types for storage, transportation, order picking and sorting.
- Sizing and dimensioning:
  - Warehouse sizing determines the storage capacity of a warehouse. There are two scenarios in modeling: (1) inventory levels are determined externally so the warehouse has no direct control over when incoming shipments will arrive and their quantities, and (2) warehouse can directly control the inventory policy.
  - Warehouse dimensioning: It translates capacity into floor space in order to assess construction and operating costs.

Table 4.8 Framework, tools and key references (Baker and Canessa 2009)

Step	Tools and key references
	Refer to literature on business and supply chain strategy literature (e.g., Christopher (2005)) and scenario planning (e.g., Sodhi (2003))
1. Define system requirement	Warehouse role framework is provided in Baker (2007a) and role checklist in Higginson and Bookbinder (2005)
	Checklists and spreadsheet, or database, models and used
2. Define and obtain data	Useful checklists appear in Rowley (2000), McGinnis and Mulaik (2000), Bodner et al. (2002), Frazelle (2002b) and Rushton et al. (2006)
	Database and spreadsheet models are used
	Activity profiling techniques are given in Frazelle (2002b)
3. Analyse data	Planning base, planning horizon and warehouse flow charts are described in Rushton et al. (2006)
4. Establish unit loads to be used	Analytic and simulation approaches are described in Roll et al. (1989)
	A wide variety of techniques are used
	Rouwenhorst et al. (2000) set out a framework of the cluster of decisions that need to be considered
5. Determine operating procedures and methods	Rushton et al. (2006) describe warehouse zoning
	Flexibility frameworks are set out in Baker (2006, 2007b)
	Spreadsheet models and decision trees tend to be used
	Heuristic, analytic and simulation methods are described in Ashayeri and Gelders (1985)
	A heuristic approach in set out in Naish and Baker (2004)
6. Consider possible equipment types and characteristics	Decision tree examples are given in Rowley (2000) and Rushton et al. (2006)
	Spreadsheet models, as well as historic performance measures, are used
7. Calculate equipment capacities and quantities	The analytic and simulation methods described by Ashayeri and Gelders (1985) are also relevant for this step
8. Define services and ancillary operations	Checklists are used by some practitioners
	CAD software is generally used by practitioners
	Outline steps and methods are provided by Mulcahy (1994), Hudock (1998) and Frazelle (2002b)
9. Prepare possible layouts	A warehouse relationship activity chart is described in Frazelle (2002b)
	Simulation software is useful at this step (e.g., see Kosfeld 1998) and is commonly used by practitioners
10. Evaluate and assess	Analytic models are also used by practitioners
	Quantitative (e.g., financial business case) and qualitative (e.g., SWOT analysis) methods are used
11. Identify the preferred design	No specific process is described in the literature

Rouwenhorst et al. (2000) provide three different angles from which a warehouse may be viewed: Processes, resources and organization. Products arriving at a warehouse subsequently are taken through a number of steps called processes.

Resources refer to all means, equipment and personnel needed to operate a warehouse. Finally, organization includes all planning and control procedures used to run the system.

Warehouse Processes

The flow of items through the warehouse can be divided in several distinct phases or processes:

- The receiving process is the first process encountered by an arriving item. Products arrive by truck or internal transport (in case of a production warehouse). At this step, the products may be checked or transformed (e.g., repacked into different storage modules) and wait for transportation to the next process.
- In the storage process, items are place in storage locations. The storage area may consist of two parts: The reserve area, where products are stored in the most economical way (bulk storage), and the forward area, where products are stored for easy retrieval by an order picker. Products in the forward area are often stored in smaller amounts in easily to access storage modules.
- Order picking refers to the retrieval of items from their storage locations and can be performed manually or (partly) automated. In succession, these items may be transported to the sorting and/or consolidation process.
- At the shipping area, orders are checked, packed and eventually loaded in trucks, trains or any other carrier.

Warehouse Resources

A number of resources can be distinguished as listed below:

- Storage unit, which products may be stored. Examples of storage units are pallets, carton boxes and plastic boxes.
- Storage system, which consist of multiple subsystems that store different types of products. Storage systems are very diverse, such as are driven in, case flow racks, shelves, push backs, just to enumerate some examples.
- The retrieval of items from the storage system can be performed manually or by means of pick equipment like forklift (e.g., single, double, triple and quad forklifts) or pallet jack.
- A computer system may be present to enable computer control of the processes by a warehouse management system.
- The material handling equipment for preparation of the retrieved items for the expedition includes sorter systems, palletizer and truck loaders.
- Personnel constitutes and important resource, since warehouse performance largely depends on their availability and commitment.

Chakravorty (2009) shows that the implementation of material handling systems involves, both human and technical factors, that interact over time, and go through three overlapping transitional stages. In the first stage, both human and technical problems exist; however, human problems dominate, and require conflict management skills to resolve. In the second stage, human problems improve, but technical problems persist, requiring formal problem-solving methods to resolve. Finally, in the third stage, both human and technical problems improve.

Warehouse Organization

Some processes require specific organizational policies:

- At the receiving process, an assignment policy determines the allocation of trucks to docks.
- At the storage process, items are transported to the storage system and are allocated to storage locations. Several storage policies exist. A dedicated storage policy prescribes a particular location for each product to be stored, whereas a random storage policy leaves the decision to the operator. In between, a class based storage system (e.g., ABC Zoning) allocates zones to specific product groups, often based upon their turnover rate. Other storage policies include correlated storage or family grouping, aimed at storing products at nearby positions, if they are often required simultaneously. If the storage system has a separate reserve area, a storage policy for the reserve area is also needed.
- At the order picking process, orders are assigned to one or more order pickers. Various control problems deserve attention. First, the total pick area may be divided into picking zones, to be served by different order pickers, through a zoning policy. Two alternative policies exist: Parallel or Sequential zoning. Second, orders are picked one by one (single order picking) or in batches (batch picking). If a batch picking policy is selected, this directly implies that the picked orders must be sorted. Then, two sorting policies exist: Pick and sort (sequentially) and sort while pick (simultaneously). Third, a routing policy may define the sequence of retrievals and the route to visit the retrieval locations.
- At the shipping process, orders and trucks are allocated to docks by a dock assignment policy.
- Finally, allocation of tasks to personnel and equipment are addressed by operator and equipment assignment policies.

Performance Management

Ellinger et al. (2005) state that the success of many firms is becoming increasingly linked to the growth, development, and retention of human capital. Accordingly, managers and leaders are being urged to promote a more people-oriented approach to management, where communication is paramount and, every employee's contribution is viewed as a significant factor in the firm's ongoing efforts to satisfy customer price, quality, and service demands. However, logistics organizations have been particularly guilty of not placing sufficient emphasis on the growth and development of personnel. This is illustrated by the difficulties that firms have in retaining truck drivers and warehouse workers who are often modestly compensated, and must perform relatively mundane and repetitive tasks.

They presented managerial coaching, which is a leadership style based on the provision of constructive feedback that is designed to get the most out of people, in other words, improve work performance by showing them that they are respected and valued. Managerial coaching occurs as part of the day-to-day relationship between employee and supervisor, and is one of the strongest retention tools in a manager's arsenal.

Managerial coaching skills include questioning, listening, giving and receiving feedback, communicating and motivating, rather than the more traditionally recognized skills and qualities of a successful leader/manager like competitiveness, being in control, solving problems, and being seen as an expert. Successful manager/coaches are also proficient delegators who are prepared to accept short-term failures, if they lead to long-term development.

Numerous prescriptive works outline methodologies for effective managerial coaching. The following list outlines behavioral practices drawn from interviews with "best of breed" manager-coaches in business and industry contexts:

- Providing observational, reflective and constructive feedback to subordinates
- Seeking feedback from subordinates about their progress on the job
- Providing resources, information, and materials for subordinates and removing external factors and obstacles that may be impeding performance
- Talking things through together to come up with options
- Stepping into subordinates' shoes to experience their perspectives and encouraging subordinates to see other perspectives
- Role-playing, personalizing learning situations with examples, and using analogies and scenarios
- Setting goals, outlining expectations and communicating their importance to subordinates
- Posing context-specific questions that encourage subordinates to think through issues themselves
- Mentally holding back and consciously not providing answers, solutions, or telling subordinates what to do; and not taking over employees' responsibilities but rather shifting them back to employees and holding them accountable

Ellinger and Keller (2005) argue that managerial coaching has enormous benefits for firms and for the individuals they employ, as people who feel valued and respected by their employers tend to be more loyal and hard-working, because they get more out of their jobs. Without coaching and feedback, effective performance is not reinforced, ineffective performance is not identified, and employees will not know if their performances are meeting the expectations of their managers or supervisor, or the requirements of the firm's customers.

Warehouse Metrics

Huertas et al. (2007) propose a set of metrics to evaluate warehouse performance as described below:

- Quality: Accuracy in storage, Accuracy in picking and Inventory
- Finance: Operational costs and Total storage costs per unit
- Cycle time: Commodity cycle time and Order cycle time
- Productivity: Labor productivity (employees/moved unit), Resource consumption (space, equipment, labor), Flow (moved units through the system in a given period), Throughput volume (moved units per day), and Productivity ratio (handled units per day/working hours per day)

It is important to mention that all concepts presented are valid for both a company owned or an outsourced warehouse operation and the decision for outsource or not, should be aligned with the company's strategic direction, financial objectives, and core function analysis. For more information on how to structure an outsource process, please refer to Greaver (1999) and Power (2008).

4.4.4.3 Distribution

A well run distribution operation is one operation that maximizes delivery asset (e.g., trucks) utilization at the same time that minimizes labor (e.g., drivers and delivery helpers) and delivers the expected customer service level (e.g., order or case fill rate).

In order to achieve the above objectives, companies need to establish the right fleet size and fleet composition, need to optimize daily vehicle routing to reduce distance traveled and drivers' working hours and also implement a performance management execution process. Each one of these three areas (fleet size, vehicle routing optimization and performance management) will be reviewed in detail below.

Fleet Size and Composition

The fleet size and composition problem was first studied by Kirby (1959), and then by Wyatt (1961), and both of them, considered a homogenous fleet to be defined, taken into consideration seasonal demand, and the fact that deliveries should be done preferable by an internal fleet, but if the demand exceeds capacity, spot fleet could be hired to fulfill the demand.

Gould (1969) brought a new view for the fleet size problem, as he considered the heterogeneous fleet for both fleet type and size.

More recently, Beaujon and Turnquist (1991) presented a research that attempts to integrate vehicle fleet sizing decisions with optimization of vehicle utilization. A model is formulated to optimize both sets of decisions simultaneously under dynamic and uncertain conditions. They showed how the expected value formulation can be approximated as a nonlinear network programming problem, and propose a procedure for solving the network approximation.

Jin and Kite-Powell (2000) state that replacement theory deals with the optimal life of an equipment. In this context, optimal life is defined as the period between the time the equipment enters service and the time when it should be replaced for economic reasons. They argue that optimal life and replacement policy are important topics in the management of capital equipment. Generally, the operating cost of a piece of capital equipment rises as its condition deteriorates over time. When the cost reaches a certain level, the long-run cost associated with investing in a new piece of equipment becomes less than that of keeping the old equipment. At that point, replacement is called for. Thus, a basic replacement analysis usually examines both the trend in operating cost and the net cost of replacement, which is defined as the difference between the cost of the new equipment and the salvage value of the old one.

List et al. (2003) proposed a formulation and a solution procedure for fleet sizing under two situations: Uncertainty in future demands, that are to be served by vehicle fleet, and the productivity of individual vehicles, reflecting uncertainty in future operating conditions.

Wu et al. (2005) describe a fleet sizing problem in the context of the truck rental industry. They examine both operational decisions (including demand allocation and empty truck repositioning) and tactical decisions (including asset procurement and sales) in a linear programming model to determine the optimal fleet size and mix. The method uses a time-space network, common to fleet management problems, but also includes capital cost decisions. A two phase solution approach is developed to solve large-scale instances of the problem using Benders decomposition in the first phase, and Lagrangian relaxation in the second phase.

Zhang and Li (2007) presented an article that analyzes multi-periodic vehicle fleet size and routing problem, and dynamic vehicle fleet size. The authors decompose the model with Dantzig–Wolf decomposition method, and derive an exact algorithm for the model based on simplex method, dynamic programming method, and branch and bound method.

Manuela (2008) studied the fleet sizing and composition problem, and compared the Bin Packing approach with an Integer linear programming model. Based on the simulation performed, the integer linear programming generated lower costs for all instances compared with the Bin Packing model, except in 1 day.

Based on the literature review, there are different approaches available to size and define the right fleet composition that companies should consider and apply, as well as they should define their fleet renew policy to ensure the fleet operation is a key enabler to deliver the customer service, cost effectively.

Vehicle Routing Optimization

Based on Toth and Vigo (2002), the vehicle routing problem (VRP) consists to determine a set of routes, each performed by a single vehicle, that starts and ends at its own depot, and ensure that all the requirements of the customers are fulfilled, all the operational constraints are satisfied, and the global distribution cost is minimized. They also presented the main components of the VRP, as summarized below:

- **Road network** – It is usually described through a graph, whose arcs represent the road sections, and whose vertices correspond to the road junctions, and to the depot and customer locations. The arcs can be directed or undirected, depending on whether they can be traversed in only one direction (for instance, because of the presence of one-way streets, typical of urban or motorway networks) or in both directions. Each arc is associated with a cost, which generally represents its length, and a travel time, which is possibly dependent on the vehicle type or on the period during which the arc is traversed.
- **Customers** – Typical characteristics of customers are:
  - Vertex of the road graph in which the customer is located
  - Amount of goods (demand), possibly of different types, which must be delivered or collected at the customer
  - Periods of the day during which the customer can be served (time windows)
  - Times required to deliver or collect the goods at the customer location (e.g., unloading or loading times, respectively), possibly dependent on the vehicle type
  - Subset of the available vehicles that can be used to serve the customer
- **Vehicle fleet** – Typical characteristics of the vehicles are:
  - Capacity of the vehicle, expressed as the maximum weight, or volume, or number of pallets, the vehicle can load
  - Possible subdivision of the vehicle into compartments, each characterized by its capacity, and by the types of goods that can be carried
  - Material handling equipments for the loading and unloading operations
  - Subset of arcs of the road graph which can be traversed by the vehicle
  - Costs associated with vehicle utilization (e.g., per distance unit, per time unit, per route, etc.)
- **Drivers** – Drivers operate the vehicles, and must satisfy several constraints laid down by union contracts and company regulations, such as working periods during the day, number and duration of breaks during service, maximum duration of driving periods, overtime, etc.

Several and often contrasting, objectives can be considered for the vehicle routing problem, and the typical ones are:

- Minimization of the global transportation cost, dependent on the global distance traveled (or on the global travel time), and on the fixed costs associated with the used vehicles (and with the corresponding drivers)
- Minimization of the number of vehicles (or drivers) required to serve all customers

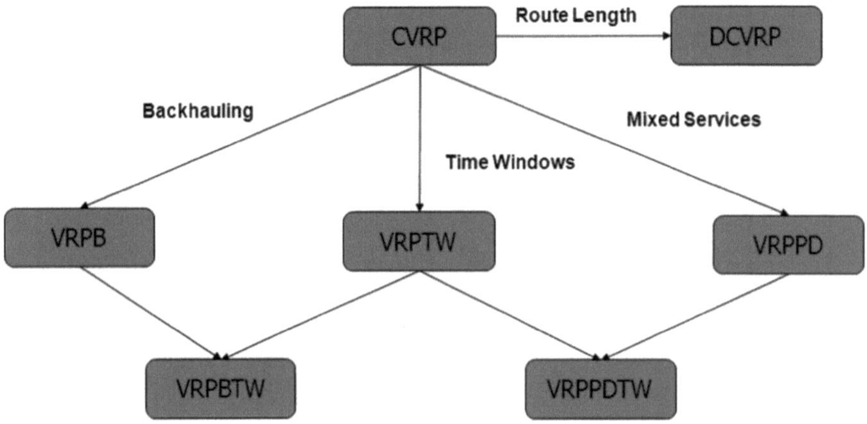

Fig. 4.20 Basic VRP problems and their interconnections (Toth and Vigo 2002)

- Balancing the routes, for travel time and vehicle load
- Minimization of the penalties associated with partial service of the customers
- Minimization of total travel time
- Maximize volume delivered per mile

Figure 4.20 provides an overview of the basic vehicle routing problems, and then, each type of problem is briefly described:

Capacitated vehicle routing problem (CVRP). The CVRP consists of finding a collection of exactly "K" simple circuits or routes (being "K" a set of identical vehicles available at the depot, each with available capacity "C") with minimum cost, defined as the sum of the costs of the arcs belonging to the routes, such that:

- Each route visits the depot vertex
- Each customer vertex is visited by exactly one route
- The sum of the demands of the vertices (customers) visited by a route does not exceed the vehicle capacity (C)

The CVRP is known to be NP-hard and generalizes the Traveling Salesman Problem (TSP), which can be defined as "given a list of cities and their pair wise distances, the objective is to find a shortest possible tour that visits each city exactly once."

One variant of CVRP is the so-called "Distance-Constrained VRP" (DCVRP), where for each route, the capacity constraint is replaced by a maximum length (or time) constraint. In particular, a nonnegative length is associated with each arc, and the total length of the arcs of each route cannot exceed the maximum route length.

Vehicle routing problem with backhauls (VRPB) is the extension of the CVRP in which the customer set is partitioned into two subsets. The first subset, "L," contains "n" Line haul customers, each requiring a given quantity of product to be delivered. The second subset, "B," contains "m" Backhaul customers, where a given quantity of inbound product must be picked up.

The VRPB consists of finding a collection of exactly "K" simple circuits or routes with minimum cost, such that:

- Each route visits the depot vertex
- Each customer vertex is visited by exactly one route
- The sum of the demands of the line haul and backhaul customers visited by a route do not exceed, separately, the vehicle capacity (C)
- In each route, all the line haul customers precede the backhaul customers, if any

The case of VRPB in which time windows are present is called "VRP with Backhauls and Time Windows" (VRPBTW).

Vehicle routing problem with time windows (VRPTW). The VRP with Time Windows is the extension of the CVRP in which capacity constraints are imposed and each customer is associated with a time interval $[a_i, b_i]$, called a time window. The time instant in which the vehicles leave the depot, the travel time for each arc, and an additional service time "s_i" for each customer should also be given.

The VRPTW consists of finding a collection of exactly "K" simple circuits or routes with minimum cost, such that:

- Each route visits the depot vertex
- Each customer vertex is visited by exactly one route
- The sum of the demands of the vertices (customers) visited by a route does not exceed the vehicle capacity (C)
- For each customer, the service starts within the time window $[a_i, b_i]$, and the vehicle stops for "s_i" time instants

Customers' time windows should be very careful defined as they increase the complexity of the routing problem to find a feasible solution, and also reduce optimization opportunities in terms of distance traveled.

In **Vehicle routing problem with pickup and delivery (VRPPD)**, each customer is associated with two quantities, "d_i" and "p_i" representing the demand of homogeneous commodities to be delivered and picked up at customer, respectively. The VRPPD consists of finding a collection of exactly "K" simple circuits or routes with minimum cost, such that:

- Each route visits the depot vertex
- Each customer vertex is visited by exactly one route
- The current load of the vehicle along the route must be nonnegative and may never exceed the vehicle capacity "C"
- Often, the origin or the destinations of the demands are common, and hence, there is no need to explicitly indicate them.

The case of VRPPD in which time windows are present is called the "Vehicle routing problem with Pickup and Deliveries and Time Windows" (VRPPDTW)

Summary of the common methods used to solve the Vehicle Routing Problem:

- Mathematical Optimization:
  - Search for optimal solution through cost minimization functions

- Classical Heuristics:
  - These methods perform a relatively limited exploration of the search space, and usually produce good quality solutions within modest computing times
  - Account for real life problem constraints
  - Examples: Constructive methods (e.g., Clarke and Wright saving algorithm, Matching-based saving algorithms, Sequential insertion heuristics); Two phase methods (e.g., Elementary clustering methods, Truncated branch and bound, Route first, cluster second methods, etc.); Improvement heuristics (e.g., Single-route improvements, Multi-route improvements, etc.)
- Meta-Heuristics:
  - General solution procedures that explore the solution space to identify good solutions, and allow deteriorating and even infeasible intermediary solutions during the search process
  - More time consuming
  - Examples: Simulated annealing, Tabu search, Ant algorithms, Neural networks etc.

It is not the goal of this thesis to develop a detail review of the solution methods available for the vehicle routing problem. For more information about each one of the above methods, please refer to Toth and Vigo (2002).

Ballou (1999) provides several principles for good routing and scheduling as detailed below:

- **Load trucks with stop volumes that are in the closest proximity to each other:** Truck routes should be formed around clusters of stops that are nearest each other, in order to minimize the interstop travel between them, which will also minimize the total travel time on the route. Figure 4.21 provides two examples of cluster routing.
- **Narrow customer time window restrictions should be avoided:** Time window restrictions usually force route sequencing away from ideal pattern. Avoid asking your customers on what time they want to receive our products.
- **Stops on different days should be arranged to produce tight clusters:** When stops are to be served during different days of the week, the stops should be segmented into separate routing and scheduling problems for each day of the week, as illustrated in Fig. 4.22.
- **Pickups should be mixed into delivery routes rather than assigned to the end of routes:** Pickups should be made, as much as possible, during the course of the deliveries to minimize the amount of path crossing that can occur when such stops are served after all deliveries are made. The extent to which this can be done will depend on the configuration of the vehicle, the size of the pickup volumes, and the degree to which they may block access to the delivery volumes inside the vehicle.
- **The most efficient routes are built using the largest vehicles available:** Ideally, using a vehicle large enough to handle all stops in one route will

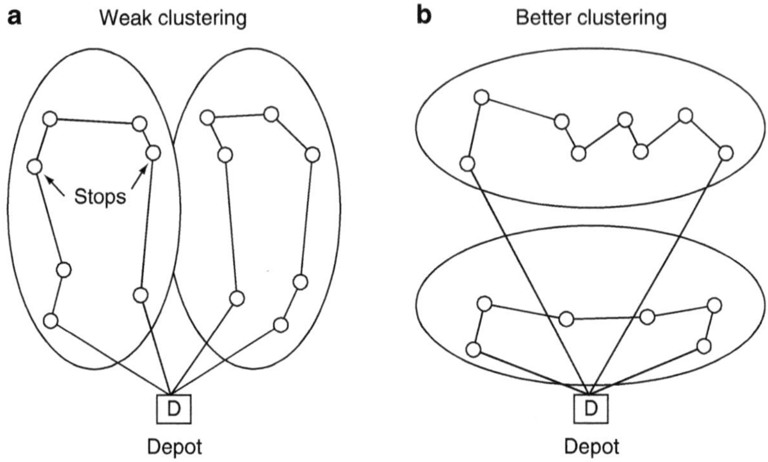

Fig. 4.21 Example of weak and good cluster routing (Ballou 1999)

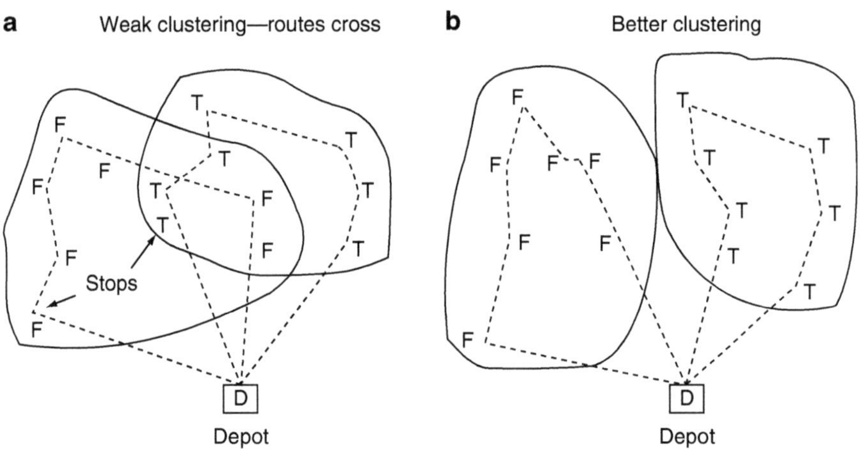

Fig. 4.22 Example of clustering routes in different days of the week (Ballou 1999)

minimize total distance, or time, traveled to serve the stops. Therefore, the largest vehicles among the multiple sizes in a fleet should be allocated first, providing that good utilization for them can be realized.

Daganzo (2010) also confirms that in any practical situation, large vehicles should be used first in order to minimize the transportation costs. He suggested using vehicle capacity as large as the highway network would allow.

- **The sequence of stops on a truck route should form a teardrop pattern:** Stops should be sequenced so that no paths of the route cross, and the route appears to have a teardrop shape.

Track and Trace Systems

Giaglis et al. (2004) state that real-time vehicle management is important in supporting supply chain execution systems, and also in minimizing related logistics risks. He argues that it has been demonstrated that a good, near-optimal, distribution plan is necessary but not sufficient for high performance distribution. This needs to be complemented by the ability to make and implement sophisticated decisions in real-time, in order to respond effectively to unforeseen events. The emergence of technologies and information systems allowing for seamless mobile and wireless connectivity between delivery vehicles and distribution facilities is paving the way for innovative approaches in addressing this requirement.

Cheung et al. (2008) developed a mathematical model for dynamic fleet management that takes into consideration dynamic data, such as vehicle locations, travel time, and incoming customer orders. The model is able to efficient re-optimize the route plan as dynamic information arrives, and it includes a genetic algorithm procedure for solving the static vehicle routing problem, and a quick heuristic procedure for dynamic updates of the vehicle routes as new data arrive.

A track and trace system can be defined as a logistics IT system that allows companies to track delivery trucks during the route in real time, and manage the operation in order to solve problems as they happen during the course of the delivery operation. These systems work integrated with routing optimization solutions, like UPS Roadnet or Descartes Roadshow, and use the route plan as the key input that drivers need to follow to reduce miles driven and missed time windows.

Example of functionalities for a track and trace solution:

- Track and manage individual performance by driver and by day
- Ability to provide closed loop system, comparing actual to schedule (hours, km) in order to improve routing planning integrity
- Ability to provide Estimated Time of Arrival (ETA) and send Advanced Shipping Notice (ASN) to customers
- Track service failures by reason code, and provide route and customer level detail, thus delivering "perfect order" report card capability
- Ability to provide delivery information to all different functional areas inside the organization (e.g., tell sell, customer service department, etc.) using web-based solution
- Allow drivers to receive planned information, and view route map and stop location on mobile device
- Ability to see planned and actual route in the same map, including playback past day routes
- Allows driver to inform in real time arrivals and departures from customers
- Allows driver to inform quantities delivered and payment information
- Allows driver to inform exceptions, sending messages (pre-defined or free messages)
- Helps fine tune customer geo-coding information (e.g., latitude and longitude)

- Have a control panel to monitor the operation in real time during the day
- Ability to create dashboard with customized set of KPIs
- Ability to track trucks during the route off line, and on line in real time, identifying exceptions in distance traveled, time stopped in a customer, arrive earlier or late, etc. All of the exceptions should be parameters defined by users.

In Fig. 4.23, the key components of a delivery track and trace system are presented. The first component is the mobile equipment (e.g., cell phone or hand held) which is used by drivers to enter delivery information as they perform the operation (e.g., arrival and departure from customers, product returns, etc.). The second component is GPS equipment that is responsible to send truck position in a pre-defined frequency (e.g., every 1 min). Both GPS information and drivers input is sent to a control operational center in the warehouse facility, where route planners can visualize exceptions to the planned route in a dashboard, as illustrated in Fig. 4.24, and work with the drivers and delivery supervisors to solve the problems or escalate them to the right person inside the organization.

Based on the author's experience on implementing track and trace systems in several different delivery operations, the typical benefits of these systems are:

- Effectively manage delivery operation during the day, managing exceptions based on standard operating procedures.
- Improve customer service through increased order fill rate, order visibility, on time delivery.
- Improve routing planning through a closed loop system that will keep an updated set of parameters (service time, street speed, etc.).

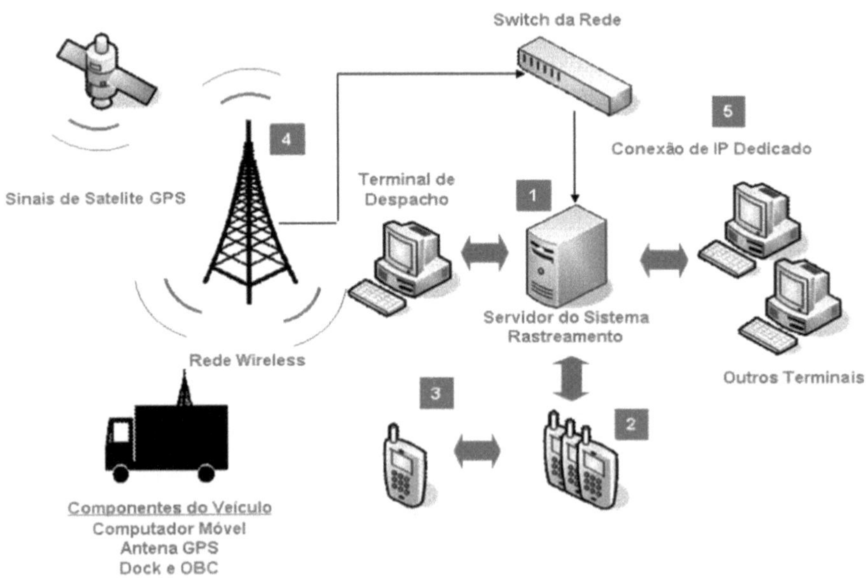

Fig. 4.23 Example of architecture for a track and trace solution

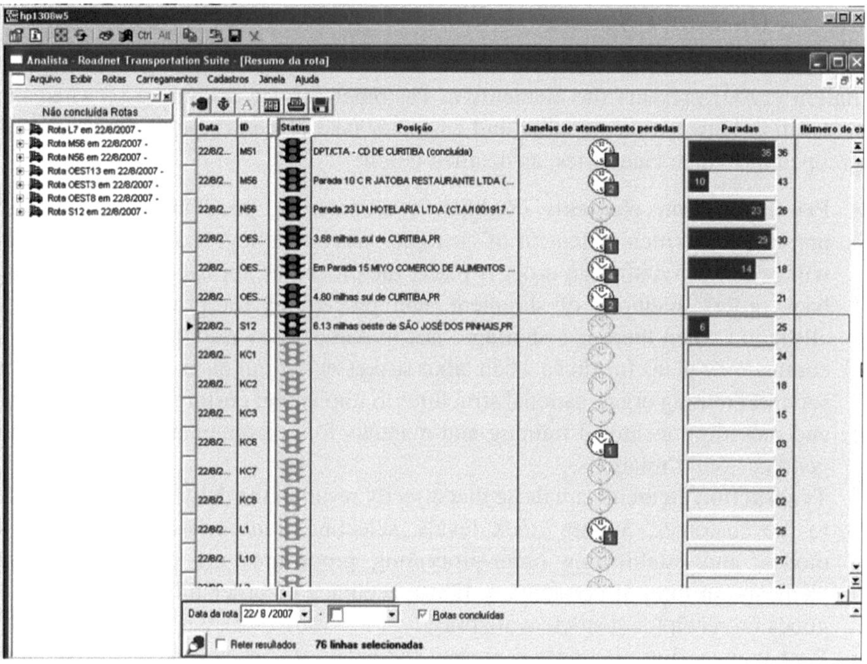

Fig. 4.24 Example of UPS mobile cast track and trace dashboard

- Improve logistics execution and reduce distribution costs (usually 10–15% reduction in km driven).
- Manage drivers' working time to reduce lawsuit possibilities due to overtime.

Performance Management

In the same way that was stated in the warehouse operation, performance management is also very important to the distribution operation as labor cost is almost 50% of the operational cost. To that end, it is proposed to apply the same managerial coaching approach, which is a leadership style based on the provision of constructive feedback that is designed to get the most out of people, in other words, improve work performance by showing them that they are respected and valued. Managerial coaching should occur as part of the day-to-day relationship between employee and distribution supervisor and is one of the strongest retention tools in a manager's arsenal. For more information about managerial coaching, please refer to Sect. 4.4.4.2.

It is important to mention that all concepts presented for distribution optimization are valid for both a company owned or an outsourced distribution operation and the decision for outsource or not, should be aligned with the company's strategic direction, financial objectives, and core function analysis.

4.4.5 *Customer Service*

Ballou (1999) presents the elements of customer service according to when the transaction between the supplier and customer take place. These elements can be grouped into three categories, as detailed below:

- **Pre-transaction elements** establish a climate for good customer service, providing a written statement of customer service policy, such as when goods will be delivered after an order is place, the procedure for handling returns and back orders, methods of shipment, and procedures for prioritizing product allocation when there is a shortage, just to name a few examples. Establishing contingency plans for times when labor strikes or natural disasters affect normal service, creating organizational structures to implement customer service policy, and providing technical training and manuals for customers also contribute to good customer relations.
- **Transaction elements** are those that directly result in the delivery of the product to the customer. Setting stock levels, selecting transportation and delivery modes, and establishing order-processing procedures are examples. These elements, in turn, affect delivery times, accuracy of order filling, condition of goods on receipt and stock availability.
- **Post-transaction elements** represent the array of services needed to support the product in the field, to protect consumers from defective products, to provide for the return of packages (e.g., returnable bottles, pallets, etc.), and to handle claims, complaints, and returns. These take place after the sale of the product, but they must be planned for in the pre-transaction and transaction stages.

Corporate customer service is the sum of all these elements because customers react to the total mix.

Segmentation is one of the key strategies to deliver a high customer service level, cost effectively, and can be defined as the process of splitting customers or potential customers within a market into different groups, or segments, within which customers share a similar level of interest in the same, or comparable, set of needs satisfied by a distinct marketing proposition.

McDonald and Dunbar (2007) provide a brief review of the predetermined approaches frequently used to perform market segmentation, and basically, there are five approaches:

- Product and services
- Demographics
- Geography
- Channel
- Psychographics

There are different types of customer service organizational structures. However, two types are predominant in place nowadays:

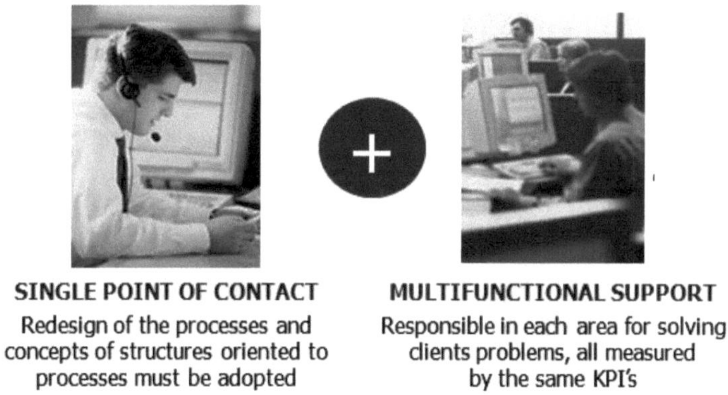

Fig. 4.25 Author's example of single point of contact structure

- **Single point of contact with support from multifunctional teams**. In this type of structure, there is one single point of contact with all customers to solve operational requirements, like changing delivery dates and frequencies, developing customized solutions, etc. The person in charge of the contact will collect customer request and will work closely with each functional area of the company to review and implement the appropriate solution. Figure 4.25 provides an example of single point of contact structure.
- **Dedicated Multifunctional Cell structure**. In this type of structure, representatives of different functional areas like Marketing, Logistics, Finance, IT, and Commercial work in the same office to serve a pre-defined customer or group of customers. This is a more specialized approach and is found in some companies of the consumer product industry, like Procter and Gamble. Figure 4.26 provides an example of multifunctional cells.

A well defined and executed customer service policy can have a deep impact in the company financial returns, as illustrated in the Fig. 4.27.

4.4.6 Senior Management Support

Bossidy and Charan (2002) state that there are seven leadership behaviors that form the building block of execution:

- Leaders should know their people and their business
- Insist on realism
- Set clear goals and priorities
- Follow through to confirm that actions and plans are executed
- Reward the "doers"
- Expand people's capabilities
- Leaders should know themselves

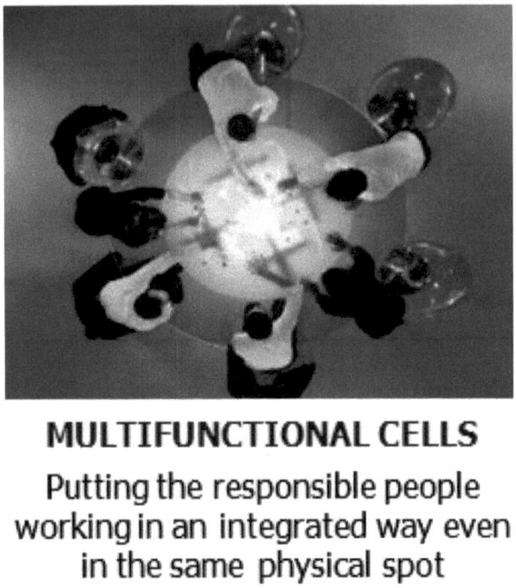

Fig. 4.26 Example of dedicated cell structure

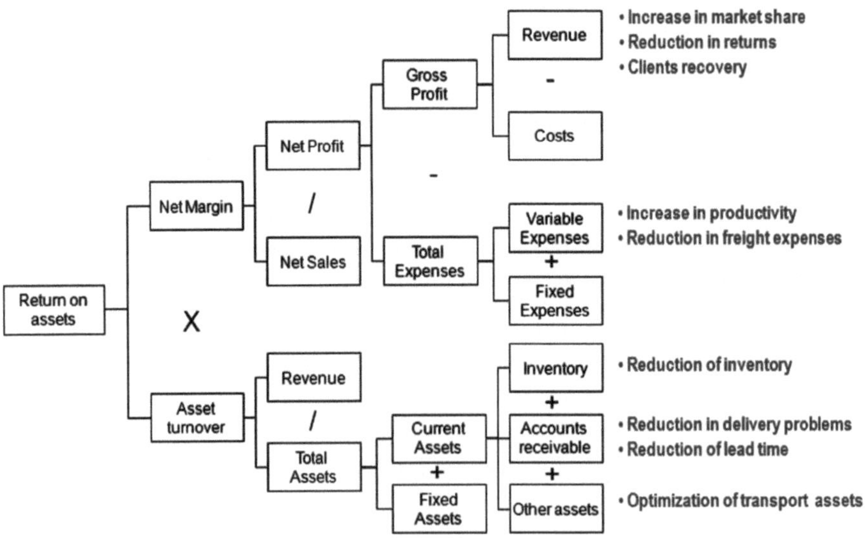

Fig. 4.27 Potential impact of a well executed customer service policy

Nalamalapu (2004) argues that leaders should have the following characteristics and behaviors:

- Admit their own mistakes
- Be able to adapt to new situations
- Be empathetic

- Leaders should sacrifice for and dedicate themselves to their missions
- Leaders should be disciplined
- Be effective communicators
- Lead by example

Fiedler's contingency theory shows the relationship between the leader's orientation or style and group performance under different situational conditions. Fiedler found that task oriented leaders were more effective in low and moderate control situations and relationship oriented managers were more effective in moderate control situations (http://www.valuebasedmanagement.net/methods_contingency_theory.html visited on 25 Sept 2010).

In order to move towards a demand driven supply and operations management, it is important to have senior management support and engagement to make the necessary changes in term of tools, processes and organizational culture, as detailed below:

- Tools – It is necessary to implement new IT systems that will share demand and supply information, first inside the organization, and then, across the supply chain with customers and suppliers to allow sense and respond to real time demand.
- Processes – As the company moves towards a pull operation, new processes and operating procedures should be developed and implemented, which will require sufficient employee training and management leadership to ensure that new processes will be engrained into the shop floor. Senior management support is crucial to provide enough resources and leadership during this transition period.
- Organizational culture – The transition from a reactive push operation to a proactive pull operation requires a change in the organizational culture, and senior management support is fundamental to allow HR design and implement required programs to embed the new culture in the organization.

Supply chain directors should also perform educational workshops with other directors and the CEO (chief executive officer) to allow them understands the differences between a "pure push" operation to a hybrid or a "pure pull" operation, depending on the objectives of the company.

4.5 Product Lifecycle Management (PLM)

In this section, it will be performed a literature review for each one of the six categories of Product Lifecycle management – New Product Forecast, Supply chain Approach for New Product, Risk Assessment and Management, Product Tracking & Visibility, Portfolio Optimization, Senior Management Support. This review allowed identify the DDSC characteristics for each category which was used to develop the five level maturity model.

4.5.1 Introduction

Based on Kahn (2005), product innovation – the development of new and improved products – is crucial to the survival and prosperity of the modern corporation. He stated that according to an American Productivity and Quality Control (APQC) benchmarking study, new products launched 3 years before the study, accounted for 27.5 of company sales, on average (American Productivity and Quality Center 2003), and product lifecycles are getting shorter with a 400% reduction over the last 50 years as a result of an accelerating pace of product innovation. However, many products do not succeed, as only 56% of businesses' new product development projects achieve their financial goals, and only 51% are launched on time.

To cope with this new scenario, product lifecycle management as the integrated, information – driven approach to all aspects of a product's life, from concept to design, manufacturing, maintenance and removal from the market, has become a strategic priority in many company's boardrooms (Teresko 2004).

For example, in the pharmaceutical industry, the development time for new drugs has almost doubled over the last 30 years, and the average drug development costs exceed US$ 800 million. Reshaping the lifecycle curve, so that, profitability starts earlier and maturity ends later is seen as a matter of survival (Daly and Kolassa 2004).

The automotive industry is another vivid example of where success or failure is strongly influenced by the company's ability to proactively manage product lifecycles (Korth 2003). Increased product complexity, greater reliance on outsourcing and a growing need for collaboration with a rapidly expanding list of business partners are the specific PLM challenges the industry faces (Teresko 2004). Furthermore, in high-tech or fashion industries, accelerated technological and design changes explain why PLM is at the forefront (Supply Chain Manager Europe 2005).

In the beverage industry, it can also be seen a huge increase in the number of SKUs commercialized in different markets and geographies, as illustrated in Fig. 4.28.

PLM confronts the need to balance fast response to changing consumer demands with competitive pressure to seek cost reductions in sourcing, manufacturing and distribution. It needs to be based on a close alignment between customer-facing functions (e.g., marketing, sales, customer service) and supply functions (e.g., purchasing, manufacturing, logistics) (Combs 2004) and (Conner 2004).

4.5.2 Proposed PLM Strategic Framework

In order to cope with this new scenario of high SKU proliferation and short product lifespan, it is proposed to apply the PLM strategic framework developed by the

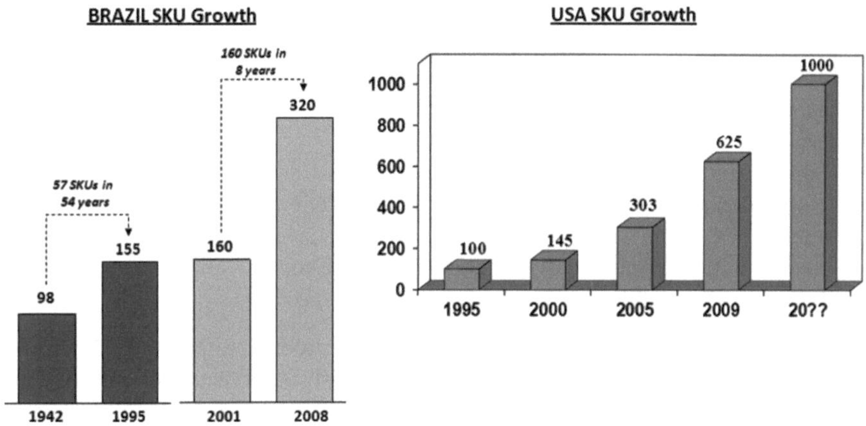

Fig. 4.28 SKU growth in one beverage industry in Brazil and USA markets

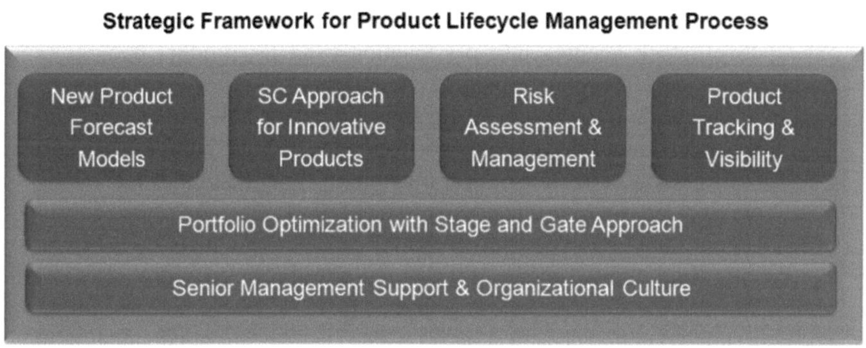

Fig. 4.29 Author's proposed PLM strategic framework

author that consists of four pillars and two foundational blocks as detailed in Fig. 4.29. Each one of the pillars and blocks will be detailed in the section below.

4.5.2.1 New Product Forecast Models

New product forecast is an important capability to guarantee product availability in the early phases of introduction in the market. However, benchmarking research shows that new product forecast accuracy is only slightly above 50% on average 1 year after the launch.

Common issues related to new product forecasting are:

- Low accuracy (high forecast error) and high variability
- Low forecast credibility inside the organization and a lot of complaints from other areas

- Limited amount of data available for analysis or to be used in the forecast process (usually, sales data is available only for regular products)
- Slow operational process to adjust or change based on market/demand signals
- Inability to fully capture market complexity, cannibalization, market penetration rate, etc.
- Limited amount of time for analysis and forecast
- Information about new product introduction usually is not provided in appropriate time
- Operational problems usually increase forecast error (raw material availability, Out-of-Stocks in the market, inventory problems, etc.)

Kahn (2006) provides direction on how new product forecasting should be performed. He states that the first step is to establish the forecasting objective, as this will clarify the purpose and intent of the forecast so that a meaningful forecast can be made. Once the forecasting objective is established, consideration is needed regarding the forecasting level, time horizon, interval and form.

- Forecasting level refers to the focal point in the corporate hierarchy where the new product forecast applies. Common levels include the stock keeping unit (SKU) level, stock keeping unit per location (SKUL), product line, strategic business unit (SBU) level, company level, and industry level.
- Forecasting time horizon refers to the time frame for how far out one should forecast. New product forecasts may correspond to a single point in the future or a series of forecasts extending out for a length of time. Examples include a 1–2 year time horizon for most fashion products, 2–5 years for most consumer products goods, and ten-plus years for pharmaceutical products.
- Forecasting time interval refers to the granularity of the new product forecast with respect to the time bucket as well as to how often the forecast might be updated. For example, a series of forecasts can be provided on a weekly, monthly, quarterly, or annual basis.
- Forecasting form refers to the unit of measure for the forecast. Typically, early new product forecasts are provided in monetary form (e.g., US dollars) and later provided in terms of unit volume for production purposes.

There are seven different types of new products, as described below:

- **New products focused on cost reductions** are products that do not have dramatic changes, but have changes that can influence consumer purchase behavior, especially when connected with implementing a new pricing policy or sustaining a cost advantage.
- **Product improvements** are product enhancements that improve the product's form or function and are often labeled as "new and improved" or "better flavor."
- **Line extensions** retain standard features of an original product (or set of products) and add unique features that the original product (or original set of products) does not have. The distinction between a product improvement and line extension is that the product improvement replaces the original product, so

Fig. 4.30 Product–Market matrix (Kahn 2006)

customers are migrated to the new product, while in the case of a line extension both the original and new products are available for purchase.

- **New market products** are when a company takes its product to a new market where the product had not been offered.
- **New uses** are original products positioned in new markets without changing or only slightly changing the original product.
- **New category entries** are products that are new to the company, but as a category, not new to the consumer.
- **New to the world products** are technological innovations that create a completely new market that previously did not exist.

Kahn (2006) provides the Product–Market Matrix in Fig. 4.30 which organizes these seven types above into two dimensions (Market and Product), and states four different strategies:

- **Market penetration strategy** has the objective to increase market share or increase product usage. Cost reductions and product improvements are characteristic of a market penetration strategy.
- **Product development strategy** derives from an objective to capitalize on existing product technology and offer more options to the customer base. In this way, the company with a more diverse product line can fend off competitors. Line extensions are characteristically associated with a product development strategy.
- **Market development strategy** stems from a desire to expand sales volume of existing products through new markets. This would include geographic expansions, including international markets and targeting new market segments within the domestic market. New uses and new market products are characteristic of a market development strategy.
- **Diversification strategy** is pursued when the company wishes to expand its business into related businesses and unrelated businesses. A company pursuing this strategy confronts complexities associated with new customer markets and new product technologies. New categories entries and new to the world products are examples of diversification strategy.

There are five categories of new product forecasting techniques, as detailed below:

- **Judgment Techniques** represent those techniques that attempt to turn experience, judgment, and intuition into formal forecasts. There are six popular techniques within this category:
  - Jury of executive opinion: Top-down forecasting technique where the forecast is arrived at through the ad hoc combination of opinions and predictions made by informed executives and experts.
  - Sales force composite: Bottom-up forecasting technique where individuals (typically salespeople) provide their forecasts. These forecasts are then aggregated to calculate a higher level forecast.
  - Scenario analysis: Analysis involving the development of scenarios to predict the future. Two types of scenario analysis could be used – Exploratory which starts in the present and moves out to the future based on current trends, and Normative which leaps out to the future and works back to determine what should be done to achieve what is expected to occur.
  - Delphi method: Technique based on subjective expert opinion gathered through several structured anonymous rounds of data collection. The objective is to capture the advantages of multiple experts in a committee, while minimizing the effects of social pressures to agree with the majority, ego pressure, and influence of a dominant individual.
  - Decision tree: Probabilistic approach to forecasting where various contingencies and their associated probability of occurring are determined.
  - Assumptions based modeling: Technique that attempts to model the behavior of the relevant market environment by breaking the market down into market drivers, and then by assuming values for these drivers, forecasts are generated.
- **Customer/Market Research Techniques** include those approaches that collect data on the customer/market and then systematically analyze these data to draw inferences on which to make forecasts. Four general classes of techniques are available:
  - Concept testing: Process by which customers (current and potential ones) evaluate a new product concept and give their opinions on whether the concept is something that they might have an interest in and would be likely to buy. The purpose of concept testing is to proof the new product concept.
  - Product use testing: Process by which customers (current and potential ones) evaluate a product's functional characteristics and performance. The purpose of product use testing is to proof the product's functionality.
  - Market testing: Process by which targeted customers evaluate the marketing plan for a new product in a market setting. The purpose of market testing is to proof the proposed marketing plan and the "final" new product.
  - Premarket testing: Procedure that uses syndicated data and primary consumer research to estimate the sales potential of new product initiatives.

- **Time Series Techniques** analyze sales data to detect historical sales patterns and construct a representative graph or formula to project sales into the future. Time series techniques used in association with new product forecasting include:
  - Trend analysis: A line is fit to a set of data. This is done either graphically or mathematically based on the actual demand as it occurs.
  - Moving average: Technique that averages only a specified number of previous sales periods.
  - Looks-like analysis: Technique that attempts to map sales of other products onto the product being forecast. Looks-like analysis is a popular technique applied to line extensions by using sales of previous product line introductions to profile sales of the new product.
  - Autoregressive Moving Average (ARMA)/Autoregressive Integrated Moving Average (ARIMA) models: A set of advanced statistical approaches to forecasting that incorporate key elements of both time series and regression model building. Three basic activities (or stages) are considered: (1) identifying the model, (2) determining the model's parameters, (3) testing/applying the model.
- **Regression Analysis Techniques** use exogenous or independent variables, and through statistical methods, develop formula correlating these with a dependent variable. Four popular techniques are used:
  - Linear regression: Statistical methodology that assesses the relation between one or more managerial variables and a dependent variable (sales), strictly assuming that these relationships are linear in nature.
  - Event modeling: Linea regression-based methodology that assesses the relation between one or more events, whether company-initiated or nonaffiliated with the company, and a dependent variable (sales).
  - Nonlinear regression: Statistical methodology that assesses the relation between one or more managerial variables and a dependent variable (sales), but these relationships are not necessarily assumed to be linear in nature.
  - Logistic regression: Statistical methodology that assesses the relation between one or more managerial variables and a binary outcome, such as purchase versus non purchase. A logistic regression model calculates the probability of an event occurring or nor occurring.
- **Other quantitative techniques** proposed by Kahn (2006) include those techniques that employ unique methodologies or represent a hybrid of time series and regression techniques. Examples of these forecasting techniques are:
  - Expert systems: Typically computer-based heuristics or rules for forecasting. These rules are determined by interviewing forecasting experts and then constructing "if-then" statements.
  - Simulation: Approach to incorporate market forces into a decision model. A typical simulation model is Monte Carlo simulation, which employs randomly generated events to drive the model and assess outcomes.

4.5.2.2 Supply Chain Approach for Innovative Products

Fisher (1997) proposes a framework to define what is the best supply chain strategy for a company's product. He argues that the first step in devising an effective supply chain strategy is to consider the nature of the demand for the products. In this case, many aspects are important, for example, product lifecycle, demand predictability, product variety, market standards for lead times and service (the percentage of demand filled from in stock goods). He suggests classifying products into "functional" or "innovative," in order to ensure a perfect match between the type of the product and the type of supply chain. Below is a summary describing the characteristics of Functional and Innovative Products:

Functional products	Innovative products
Product do not change much over time	Great variety of products
Have stable and predictable demand	Increase unpredictability (volatile demand)
Long life cycles	Short life cycles
Lower potential growth	Higher potential growth

The next step is to decide whether the company's supply chain should be "**Physically Efficient**" or "**Responsive to the Market,**" as described in the Table 4.9.

After determining the nature of the product demand (e.g., functional and innovative products) and the supply chain priorities (e.g., responsive or efficient), managers can employ a matrix to formulate the ideal supply chain strategy. Fisher proposes to plot the nature of the demand for each of the product families and its supply chain priorities, as illustrated in Fig. 4.31, in order to allow identify whether the process used for supplying products is well matched to the product type, which means, an efficient process for functional products and a responsive process for innovative products.

Table 4.9 Physically efficient vs. market responsive supply chains (Fisher 1997)

	Physically efficient process	Market responsive process
Primary purpose	Supply predictable demand efficiently at the lowest possible cost	Respond quickly to unpredictable demand in order to minimize stock outs and obsolete inventory
Manufacturing focus	Maintain high average utilization rate (reduce setups)	Deploy excess buffer capacity
Inventory strategy	Generate high turns and minimize inventory throughout the chain	Deploy significant buffer stocks or end products in the chain
Lead time focus	Shorten lead time as long as it does not increase cost	Invest aggressively in ways to reduce lead time
Approach to choosing suppliers	Select primarily for cost and quality criteria	Select primarily for speed, flexibility, and quality
Product-design strategy	Maximize performance and minimize cost	Try to postpone product differentiation for as long as possible in the supply chain

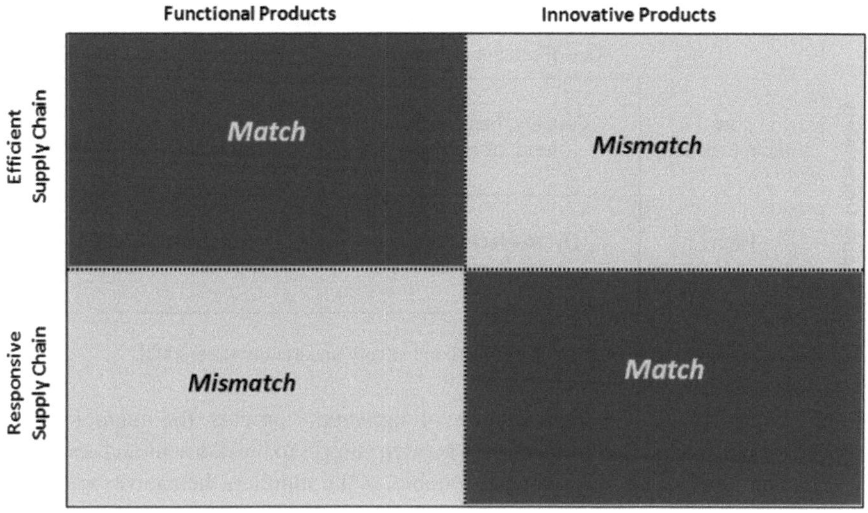

Fig. 4.31 Approach for matching supply chains with products (Fisher 1997)

When thinking about innovative products, uncertainty about demand is intrinsic to this type of products and should be accept by companies. Once a company has accepted the uncertainty of demand, it can employ three coordinated strategies to manage that uncertainty:

- First, company can continue to strive to reduce uncertainty by, for example, finding sources of new data that can serve as leading indicators or by having different products share common components as much as possible, so that, the demand for components becomes more predictable.
- Second, company can avoid uncertainty by cutting lead times and increasing the supply chains' flexibility so that it can product to order or at least manufacture the product at a time closer to when demand materializes and can be accurately forecast.
- Third, once uncertainty has been reduced or avoided as much as possible, it can hedge against the remaining residual uncertainty with buffers of inventory or excess capacity.

Lee (2002) expanded the framework developed by Fisher (1997) and introduced the concept of supply uncertainty. He argues that uncertainties around the supply side of the product are equally important drivers for the right supply chain strategy.

A stable supply process is one where the manufacturing process and the underlying technology are mature and the supply base is well established. On the other hand, an "evolving" supply process is where the manufacturing process and the underlying technology are still under early development and are rapidly changing, and as a result, the supply base may be limited in both size and experience.

In a stable supply process, manufacturing complexity tends to be low or manageable. Stable manufacturing processes tend to be highly automated, and long-term

Supply Uncertainty	Demand Uncertainty: Low (Functional Products)	Demand Uncertainty: High (Innovation Products)
Low (Stable Process)	Grocery, basic apparel, food, oil and gas	Fashion apparel, computers, pop music
High (Evolving Process)	Hydro-electric power, some food produce	Telecom, high-end computers, semiconductor

Fig. 4.32 Examples of products with demand and supply uncertainty (Lee 2002)

supply contracts are prevalent. In an evolving supply process, the manufacturing process requires a lot of fine-tuning and is often subject to breakdowns and uncertain yields. The supply base may not be as reliable, as the suppliers themselves are going through process innovations.

Figure 4.32 provides some examples of products that have different demand and supply uncertainties. Lee's advocates that it is more challenging to operate a supply chain that is in the right column of the figure below than in the left column, and similarly, it is more challenging to operate a supply chain that is in the lower row of the figure below than in the upper row. He argues that before setting up a supply chain strategy, it is necessary to understand the sources of the underlying uncertainties and explore ways to reduce these uncertainties.

Lee (2002) suggests strategies to reduce demand and supply uncertainties, as detailed below:

- **Demand uncertainty**: Information sharing and tight coordination can allow companies to regain control of supply chain efficiency. Sharing of demand information and synchronized planning across the supply chain are crucial for this objective.
- **Supply uncertainty**: Free exchanges of information – starting with the product development stage and continuing with the mature and end-of-life phases of the product life cycle – have been found to be highly effective in reducing the risks of supplier failure.

Sharing product rollover plans with suppliers is a key way to manage the risks of product transitions. Another way to reduce supply uncertainty downstream is to collaborate in the early phases of product design.

4.5.2.3 Risk Assessment and Management

Kahn (2005) states that a well planned and executed risk assessment and analysis should be part of any significant product development project. Risk assessment and analysis serve as the project manager's binoculars to see into the future. By spending time and money early to examine what might go wrong and how it might happen, managers can make effective plans to avoid those problems and/or minimize their impact, while staying on schedule and budget for the launch. He also states that project manager should apply a robust risk analysis approach to identify general potential problems in the early phase of the new product development, and later, in the development process, should use Failure Mode and Effects Analysis (FMEA), in order to have a more detailed view of each part of the design and ensure that it is aligned with the plan.

The author developed and proposes to apply the risk assessment detailed below as part of the analysis in the early phases of new product launch:

- Consider ten standard risk factors to be evaluated by project manager. These factors were identified based on the author's experience and should include Level of innovation, product coverage in the market, forecast complexity, time to market, product shelf life, manufacturing complexity, raw material characteristics, storage and operational requirements, distribution complexity, expected profitability, etc. (detail description of each factor is available in Table 4.10).
- Use a pre-defined scale to weight each risk factor. A weight of 5 means a high risk, a 3 means medium risk, and a 1 means a low risk.
- For each risk factor, there is a pre-defined functional area inside the company that will be responsible to give the score for the factor in one particular project.
- Maximum score is calculated as the Σ (Highest score for each factor × Weight for each factor).
- Final risk score for a specific new product launch is calculated as the Sum of the weighted scores for each factor/Maximum score.
- Based on risk assessment results, identify the level of risk for the project as defined below:
  - Less than 50% – Launch with LOW risk
  - Between 50 and 70% – Launch with MEDIUM risk
  - Greater than 70% – Launch with HIGH risk
- Project team should identify the areas with more critical potential risks and develop mitigating actions to reduce exposure.
- Apply Failure Mode and Effects Analysis (FMEA) technique for each project with high risk during the project development. FMEA is a 6-Sigma technique that can be applied to new product development and will allow the project team to systematically examine each important aspect of the product launch to identify, prioritize and set mitigation strategies for potential gaps/weaknesses in the new product introduction, avoiding potential future problems.

4.5.2.4 Product Tracking and Visibility

One of the key characteristics of demand driven supply chains is the ability to read and sense the demand, and have an agile supply network capable of reacting very

Table 4.10 Author's proposed list of risk factors

Risk ID	Risk factor	Responsible	Standard score	Definition
1	Level of innovation	Marketing	3	Complete new product for a new market/ channel
			2	New product for an existing market or Product for a new market
			1	Product extension or improvement
2	Product coverage in the market	Marketing	3	Product is planned to cover all channels in the market
			2	Product is planned to cover a few channels in the market
			1	Product is planned to cover few customers
3	Times to market	Marketing	3	Information about launch is available less than 20 days to launch the product in the market
			2	Information about launch is available less than 60 days to launch the product in the market
			1	Information about launch is available less than 120 days to launch the product in the market
4	Forecast complexity	Sales	3	Forecast requires to use more advanced models (conjoint, multiple regressions, ATAR, etc.)
			2	Similar products were already launched and we can use their historical information to forecast (Looks like, Curve Fitting, etc.)
			1	Some sales data is available to forecast with basic models (moving average, exponential smoothing, etc.)
5	End product shelf life	Quality	3	Shelf life $\leq$60 days
			2	Shelf life between 60 and 120 days
			1	Shelf life greater than 120 days
6	Manufacturing complexity	Manufacturing	3	Product is produced external in an Co packer or in a facility from another Bottler
			2	Product is produced internally in the franchise
			1	Product is produced in most of the sites of the franchise
7	Unique raw materials (top 5)	Manufacturing	3	Product with unique ingredients
			2	Ingredients are used by some other products in equal or major volume
			1	Ingredients are common with several other formulations in equal or greater volumes
8	Storage and operational requirements	Logistics	3	Product requires specific storage, handling and/or picking requirements
			2	Product requires only specific storage requirements (e.g., racks)

(*continued*)

Table 4.10 (continued)

Risk ID	Risk factor	Responsible	Standard score	Definition
			1	Product does not require any specific operational requirement
			3	Product requires special storage characteristics inside the truck and special handling attention during the delivery process
			2	Product requires special handling attention during the delivery process
9	Distribution complexity	Logistics	1	Product does not require any specific requirement during delivery
			3	Expected profitability is higher than average
			2	Expected profitability is equal to company average
10	Expected profitability	Finance	1	Expected profitability is slightly lower than average

fast to fulfill market demand. In order to achieve this desired state, it is critical to establish processes and implement tools that will allow having visibility of demand levels in the market and also tracking product performance in the market to compare planed vs. actual performance in different dimensions.

Based on the author's experience, it is suggested to consider the following examples of tracking:

- Total volume evolution
- Volume mix by market channel
- Repeat purchase frequency evolution
- Repeat purchase frequency by market channel
- Price compliance

When moving to a more mature demand driven state, companies can benefit from tools like CPFR, VMI, Inventory visibility, which provide actual demand and inventory information between customers and the company increasing agility and flexibility to adjust production and distribution based on real demand signals.

4.5.2.5 Portfolio Optimization with Stage and Gate Approach

Portfolio optimization is an analytical process used to determine the merits of adding, retaining, or deleting items from the product portfolio of a business. It can be summarized as a way to make sure that a company does not keep what they don't need. Portfolio optimization is important as all SKUs become liabilities to an organization at some point in their lifecycles. The key is to establish repeatable processes to identify when that point is reached and execute plans to capture as much revenue and profit as possible, before discontinuing each item. The aim is to separate underperforming SKUs from those that contribute significantly to the business results.

Portfolio optimization is used in the following situations:

- To make room for Innovation: Convince company's customers to add new products to the category and find a place for a new item on the crowded shelf in their stores.
- To increase competitiveness: Straggling competitive SKUs can take up space and bring down the profitability and productivity of the entire category, lowering the category's utility in the minds of buyers and putting the entire category's space allocation at risk.

A proactive portfolio optimization exercise not only introduces the opportunity to eliminate competitive items, but also can enhance the overall vitality of the category.

- To correct out of stock issues: Large size and high volume items can suffer from out-of-stock issues which lower potential sales. Portfolio optimization is critical to free up space for the items going out-of-stock. The high velocity items will benefit through increased profitability due to customers not being forced to go to other locations to find the high demand item.

Kahn (2005) states that a systematic new product framework such as a Stage – Gate™ process is the solution that many companies have turned to in order to overcome the deficiencies that plague their new product programs. Stage – Gate™ frameworks are simply roadmaps for driving new products from idea to launch successfully and efficiently. He argues that about 68% of U.S. product developers have adopted Stage – Gate™ frameworks, according to the 1997 PDMA best practice study. The 2003 APQC benchmarking study reveals that 73% of businesses employ such a framework and identified a stage-and-gate process as the strongest best practice, employed by almost every best performing business.

Follow below a typical five Stage-Gate™ process:

- **Stage 1 – Scoping**: A quick investigation and sculpting of the project. This first and inexpensive homework stage has the objective of determining the project's technical and marketplace merits. Stage 1 involves desk research or detective work – little or no primary research is done here. Prescribed activities include preliminary market, technical assessment, and business assessment.
- **Stage 2 – Build the business case**: The detailed homework and upfront investigation work. This second homework stage includes actions, such as a detailed market analysis, user needs and wants studies to build "voice of the customer," competitive benchmarking, concept testing, detailed technical assessment, source of supply assessment, and a detailed financial and business analysis. The result is a **business case** – a defined product, a business justification, and a detailed plan of action for the next stages.

- **Stage 3 – Development**: The actual design and development of the new product. Stage 3 witnesses the implementation of the development plan and the physical development of the product. Lab tests, in-house tests, or alpha tests ensure that the product meets requirements under controlled conditions. The "deliverable" at the end of stage 3 is a prototype product that has been lab tested and partially tested with the customer.
- **Stage 4 – Testing and validation**: The verification and validation of the proposed new product, its marketing and production. This stage tests and validates the entire viability of the project: the product itself via customer tests, beta tests, or field trials; the production process via trial or limited production runs; customer acceptance by way of a test market or a trial sell. Also, the financial justification required prior to full launch is obtained.
- **Stage 5 – Launch**: Full commercialization of the product – the beginning of full production and commercial launch and selling. The post-launch plan – monitoring and fixing – is implemented, along with early elements of the life-cycle plan (new variants and releases, continuous improvements).

Twelve to eighteen months after the launch, the post launch review occurs. The performance of the project versus expectations is assessed, along with reasons why events occurred and what lessons were learned. The team is disbanded and recognized and the project team is terminated.

Preceding each stage is a gate. These are the quality control checkpoint in the process, opening the door for the project to proceed to the next stage. Here, the project team meets with senior management, the gatekeepers, seeking approval and resources for their project. Each "go/kill" specifies deliverables (what the project team must deliver for that gate review), criteria for a go decision (for example, a scorecard as outlined above, upon which the "go/ill" and prioritization decisions are based), and outputs (an action plan for the next stage, and resources approved).

Two types of gates are possible:

- The first type of gate is called a *rigid gate* where all necessary criteria must be passed in order to continue the project. Indeed, it is possible in the case of a rigid gate that the failure to reach even a single threshold could derail a new project.
- The second type of gate is called a *flexible or permeable gate*, which allows a limited number of tasks, frequently those with long lead times, to move forward to the next stage without having to pass all criteria.

Kahn (2005) presents the top six gate criteria used by gate, as detailed in the Table 4.11 (numbers represent the percentage of firms in sample reporting to use each criterion).

Benefits of a Stage-Gate™ Approach:

- Puts discipline into a somewhat ad-hoc, chaotic process
- A visible process – known and understood by all stakeholders inside the operation
- A roadmap for the Project Leader – defines his/her duties and deliverables

Table 4.11 Criteria used by Gate, Kahn (2005)

Gate 1	Gate 2	Gate 3
Market potential (86%)	Technical feasibility (85%)	Product performance (86%)
Strategic fit (85%)	Sales objectives (81%)	Sales objectives (79%)
Technical feasibility (84%)	Product performance (79%)	Quality objectives (78%)
Sales objectives (81%)	Product advantage (77%)	Profit potential (75%)
Product advantage (79%)	Strategic fit (76%)	Customer acceptance (73%)
Profit potential (71%)	ROI (73%)	Product advantage (72%)

- Forces more attention to quality of execution – the Gates become quality control check points for product launch/retire
- Makes for a complete process – no critical errors of omission; no missing steps
- Multifunctional – inputs from all sides involved in the launching/retiring processes
- A faster process via parallel processing

Practical lessons to improve gate decisions:

- Gatekeeper team should consist of a different set of individuals than those actually conducting the product launch/retirement. (Individuals that spend months or years working on a project tend to get emotionally attached to it, making it difficult to view progress objectively.)
- Have clearly defined criteria (marketing, supply chain, commercial and financial) that must be met both overall and at each stage.
- Communicate Gate decisions to the entire organization to increase awareness and commitment. (Have a standard "meeting notes" format to document all decisions.)
- Have project continuation/termination decisions made by a cross-functional team rather than one person. Team generally makes better decisions than individuals acting alone.
- More innovative new products should require a higher level of tracking and monitoring (e.g., more stringent criteria, consider adding extra review points, etc.)

4.5.2.6 Senior Management Support and Organizational Culture

Cooper et al. (2004) performed a benchmark study to identify best practices in new product development with 105 companies from different industries and business segments, and one of the key findings were the importance of senior management support and commitment to develop a innovative culture inside the organization (79% of best performers have this practices implemented in their organizations), and also 65% provide strong support and empowerment to the team members, as detailed in the Fig. 4.33.

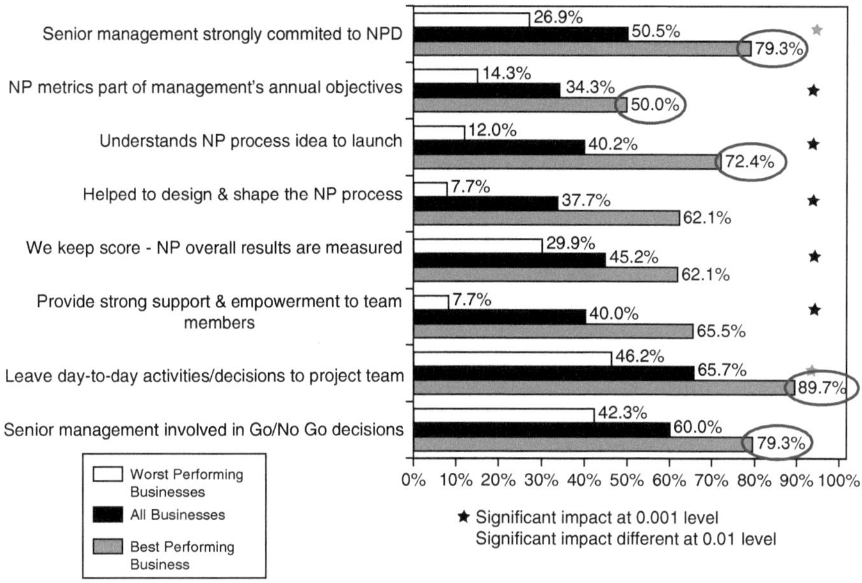

Fig. 4.33 Cooper et al. (2004) study shows the percentage of businesses where senior management demonstrates commitment to new product development

Another finding from the benchmark study is that an innovative climate inside the organization is seen as one of the most important drivers of successful product development. Twelve elements were identified and split into two main factors or themes – the first is the general climate and the second centers on specific actions and programs to promote a positive climate. The graphics in Figs. 4.34 and 4.35 present each one of the elements.

Key points for senior management attention based on author's experience:

- Provide proper direction and leadership: Senior management should foster a climate of open communication, high motivation, involvement with the project launch or retirement, and willingness from each functional area to cooperate with the project team.
- Foster an innovative culture: To really succeed with new products, senior management should align annual objectives, rewards and recognitions programs and performance management in a way that motivates and align the whole organization towards proposing and implementing innovative ideas, services and products.
- Create a culture of "don't be afraid to try": It is very important that all employees feel that they will not be punished if they try solutions "out of the box" to help the organization increase sales volume and profitability.
- Have clear metrics to support decision making: Establish a PLM dashboard to track performance of both new product introduction and product retirement aligned with business plan goals.

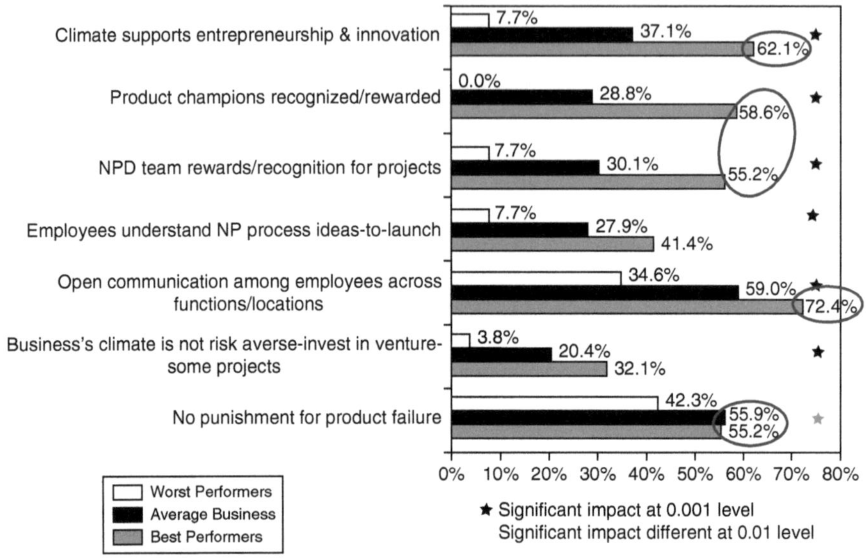

Fig. 4.34 Cooper et al. (2004) study shows the percentage of businesses that have each element of a positive climate for innovation

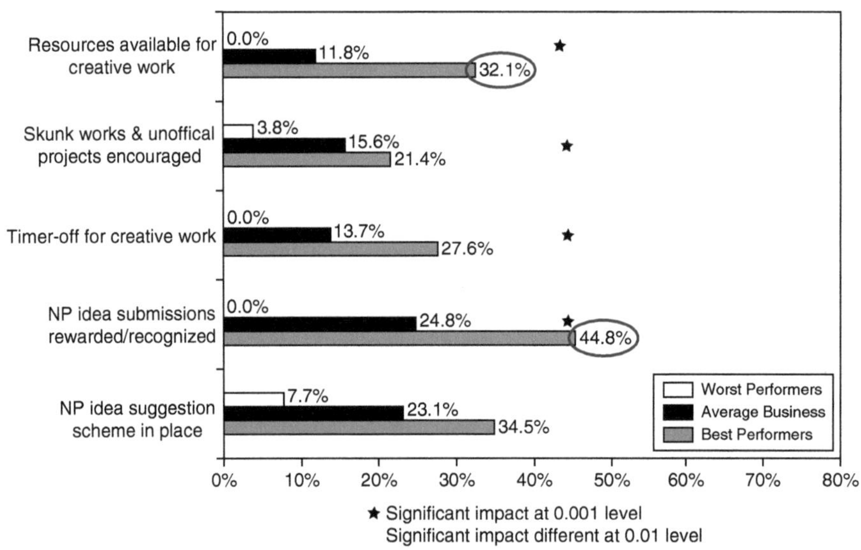

Fig. 4.35 Cooper et al. (2004) study shows the percentage of businesses with specific actions and programs to promote positive climate for innovation

References

American Productivity and Quality Center (2003) Improving new product development performance and practices. American Productivity and Quality Center, Houston

Armstrong J (2001) Principles of forecasting: a handbook for researchers and practitioners. Kluwer Academic, Norwell

Ayers J (2006) Handbook of supply chain management, 2nd edn. Auerbach, Boca Raton

Baker P, Canessa M (2009) Warehouse design: a structured approach. Eur J Oper Res 193: 425–436

Ballou R (1999) Business logistics management, 4th edn. Prentice Hall, Upper Saddle River

Beaujon G, Turnquist M (1991) A model for fleet sizing and vehicle allocation. Transp Sci 25 (01):19–45

Bossidy L, Charan R (2002) Execution: the discipline of getting things done. Crown Business, New York

Bowersox D, Lahowchich N (2008) Start pulling your chain: leading responsive supply chain transformation. OGI Enterprises, Pearland

Chakravorty S (2009) Improving distribution operations: implementation of material handling systems. Int J Prod Econ 122(1):89–106

Chatfield C (2004) The analysis of time series: an introduction, 6th edn. Chapman and Hall/CRC, Boca Raton

Cheung B, Choy K, Li C, Shi W, Tang J (2008) Dynamic routing model and solution methods for fleet management with mobile technologies. Int J Prod Econ 113:694–705

Combs L (2004) The right channel at the right time. Ind Manage 46:8–16

Conner M (2004) The supply chain's role in leveraging PLM. Supply Chain Manage Rev 8:36–43

Cooper RG, Scott JE, Elko JK (2004) Benchmarking best NPD practices. Res Technol Manage 47 (3):50–59

Croxton K, Lambert D, Garcia-Dastugue J (2002) The demand management process. Int J Logist Manage 13(2):51–66

Daganzo C (2010) Logistics systems analysis, 4th edn. Springer, Berlin

Daly R, Kolassa M (2004) Start earlier, sell more, sell longer. Pharm Exec 8:30–38

Danese P (2004) Beyond vendor managed inventory: the Glaxosmithkline case. Supply Chain Forum Int J 5:32–39

De Toni AF, Zamolo E (2005) From a traditional replenishment system to vendor-managed inventory: a case study from the household electrical appliances sector. Int J Prod Econ 96:63–79

Ellinger A, Ellinger A, Keller S (2005) Supervisory coaching in a logistics context. Int J Phys Distrib Logist Manage 35(9):620–636

Fisher M (1997) What is the right supply chain for your product. Harvard Bus Rev 75(2):105–116

Giaglis G, Minis I, Tatarakis A, Zeimpekis V (2004) Minimizing logistics risk through real-time vehicle routing and mobile technologies: research to date and future trends. Int J Phys Distrib Logist Manage 34(9):749–764

Gould J (1969) The size and composition of a road transport fleet. Oper Res Soc 20(5):81–92

Greaver M (1999) Strategic outsourcing: a structured approach to outsourcing decisions and initiatives. AMACOM, New York

Gu J, Goetschalckx M, McGinnis L (2010) Research on warehouse design and performance evaluation: a comprehensive review. Eur J Oper Res 203:539–549

Hall C (2001) What is VMI? Enterprise Data Manage

Harrison T (2003) The practice of supply chain management: where theory and application converge. International Series in Operations Research and Management Science. Springer, New York

Huertas J, Ramírez J, Salazar F (2007) Layout evaluation of large capacity warehouses, vol 25, N. 7/8. Emerald, Bingley, pp 259–270

ILOS Institute (2009) Pesquisa de Planejamento da Demanda e S&OP. Geneva, ILO. http://www.ILOS.com.br

Jin D, Kite-Powell H (2000) Optimal fleet utilization and replacement. Transp Res E 36:3–20
Kahn K (2005) The PDMA handbook of new product development. Wiley, New York
Kahn K (2006) New product forecasting: an applied approach. M.E. Sharpe, Armonk
Kellogg Graduate School of Management (2000) Working Paper on Vendor Managed Inventories
Kirby D (1959) Is your fleet the right size? Oper Res Soc 10(04):252
Korth K (2003) Learning to compete in the current environment. Automot Design Prod
Lambert D (2008) Supply chain management: processes, partnerships, performance. Supply Chain Management Institute, Sarasota
Lapide L (2003) Organizing the forecasting department. J Bus Forecast Methods Syst, p 20–21
Lapide L (2004) Sales and operations planning part I: the process. J Bus Forecasting Meth Syst 23(3):17–19
Lapide L (2005) Sales and operations planning part III: a diagnostic model. J Bus Forecasting 24:13–16
Lee H (2002) Aligning supply chain strategies with product uncertainties. Calif Manage Rev 44(3):104–119
List G, Wood B, Nozick L, Turnquist M, Jones D, Kjeldgaard E, Lawton C (2003) Robust optimization for fleet planning under uncertainty. Transp Res E 39:209–227
Lockamy A, McCormack K (2004) Linking SCOR planning practice to supply chain performance: an exploratory study. Int J Oper Prod Manage 24(12):1192–1218
Makridakis S (1998) Forecasting: methods and applications, 3rd edn. Wiley, New York
Manuela C (2008) Dimensionando uma Frota Heterogênea de Veículos Ótima em Tamanho e Composição. Dissertação de Mestrado, PUC/RJ
McDonald M, Dunbar I (2007) Market segmentation: how to do it, how to profit from it. Elsevier, Oxford
Mentzer J (1999) The impact of Forecasting on Return on Shareholder's Value. The Journal of Business Forecasting, p 8–10
Mentzer T, Cox J (1984) A model of the determinants of achieved forecast accuracy. J Bus Logist 5(2):143–155
Mentzer T, Davis D (2007) Organization factors in sales forecasting management. Int J Forecasting 23(3):475–495
Mentzer T, Moon M (2005) Sales forecasting management: a demand management approach, 2nd edn. Sage, Thousand Oaks
Moon M, Mentzer J, Smith C, Garver M (1998) Seven keys to better forecast. Bus Horiz 41(5):44–52
Nahmias S (2005) Production and operations analysis. McGraw-Hill Irwin, Boston
Nalamalapu A (2004) Leaders can learn much from Gandhi. Leadersh Action 24
Oliveira P Jr (2004) Impact of demand forecasting inaccuracy on the supply chain. a case study in the beverage industry. Master Science Dissertation Thesis, PUC-RJ
Oliver Wight White Paper Series (2005) CPFR: collaborative planning, forecasting and replenishment
Power M (2008) The outsource handbook: how to implement a successful outsourcing process. Kogan Page, London
Rouwenhorst B, Reuter B, Stockrahm V, Van Houtum G, Mantel R, Zijm W (2000) Warehouse design and control: framework and literature review. Eur J Oper Res 122:515–533
Stewart G (1997) Supply-chain operations reference model (SCOR): the first cross industry framework for integrated supply-chain management. Logist Inf Manage 10(2):62–67
Taylor DH (2000) Demand amplification: has it got us beat? Int J Phys Distrib Logist Manage 30(6):515–533
Teresko J (2004) The PLM revolution. Ind Week 253:32–36
Toth P, Vigo D (2002) The vehicle routing problem. SIAM monographs on Discrete Mathematics and Applications. SIAM, Philadelphia
Wallace TF (2004) Sales and operations planning: the "how to" handbook, 2nd edn. T F Wallace, Cincinnati

Waller M, Johnson M, Davis T (1999) Vendor-managed inventory in the retail supply chain. J Bus Logist 20:183–203
Watson E (2005) Nestle switches to vendor managed inventory with Tesco. Food Manuf 80:20
Wu P, Hartman J, Wilson G (2005) An integrated model and solution approach for fleet sizing with heterogeneous assets. Transp Sci 39(01):87–103
Wyatt JK (1961) Optimal fleet size. Oper Res Soc 12:186–187
Zhang Y, Li J (2007) Dynamic optimal model of vehicle fleet size and exact algorithm. Syst Eng Theory Pract 27(02):83–91

Chapter 5
Proposed Demand Driven Supply Chain Model

5.1 Introduction

Based on the three demand driven components (e.g., demand management, supply and operations management and product lifecycle management) and the categories within each component, identified in the previous chapter, five level maturity model was developed to be used as the basis to perform assessment of an organization current and desired future states.

It is proposed the following levels for the maturity model:

- **Level 1 – Basic Push Operation**

  In this level, only some of the foundations of a good push operation are in place, but the organization does not have all of them well implemented or misses important ones.

- **Level 2 – Optimized Push Operation**

  In this level, all foundations of a good push operation are in place and the organization captures benefits from the good execution of push principles.

- **Level 3 – Hybrid Push–Pull Operation**

  In this level, the organization starts to move from a pure Push system to a hybrid Push–Pull system, through implementation of some of the demand driven concepts.

- **Level 4 – Advanced Demand Driven (Pull) Operation**

  In this level, the organization had already implemented most of the demand driven concepts and captures benefits from fulfilling customer demand in a cost effective way.

This chapter describes the characteristics of each level inside the DDSC maturity model.

P. Mendes, *Demand Driven Supply Chain*,
DOI 10.1007/978-3-642-19992-9_5,

- **Level 5 – Optimized Demand Driven (Pull) Operation**

  In this level, the organization has not only implemented the demand driven concepts internally, but also expands them to the whole supply chain where it operates, and experienced proven financial and service improvements.

5.2 Demand Driven Supply Chain Maturity Model

The detailed Demand Driven Maturity Model is presented below:

5.2.1 Component: Demand Management

5.2.1.1 Category: Statistical Forecast

Level 1

- No statistical forecast methods or only very basic models (e.g., moving average) are used to plan business volume.
- Forecast is usually generated based on management experience and/or annual volume target growth.
- No formal demand planning organizational structure in place or planners are only part time assigned to the forecast function.
- Lack of right skill set for demand planners inside the organization.
- Demand planners and forecasts have low credibility from other functional areas inside the organization.
- Performance of demand planning and forecast are not tied to compensation and rewards.
- Low or no senior management support to demand planning function.
- No forecast tool is in place to automate statistical forecast process.
- No standard metrics are used to measure forecast accuracy, identify improvement opportunities and communicate performance to the entire organization.

Level 2

- Statistical forecast methods (e.g., Exponential Smoothing, Box-Jenkins, Holt and Holt-Winters) are used to plan business volume for short term period (1 week to 4 months). Combined forecast methods are also used to improve forecast accuracy.
- Linear regression and econometric methods are used to plan business volume for long term forecast (12 months to 5 years).
- Forecasts are presented using prediction intervals and scenarios to cope with uncertainty.

- Formal planning organizational structure is in place with clear roles and responsibilities. There is updated job description for demand planners and managers.
- Demand planners have right skill set (quantitative, computer, interpersonal skills and process management).
- Planners and generated forecasts have high credibility inside the organization and are recognized as an added value function.
- Performance of planners and forecast are tied to compensation and rewards.
- Senior management understands, support and value demand planning function.
- Forecast tools are in place for both short and long term forecast, and are used to automate statistical forecast process, increasing planners capability to simulate different models.
- Standard key performance indicators [e.g., Forecast accuracy, forecast quality, MAPE, Mean Absolute Deviation (MAD), Mean Squared Deviation (MSE), etc.] are used to measure forecast results, identify improvement opportunities and communicate performance to the entire organization.

Level 3

- Customers and channels are analyzed and clustered to identify and apply the best demand planning method for each cluster.
- Based on demand variability and sales volume, planners understand SKU profile and apply appropriate forecast methods (same as in level 2) for SKUs with low variability, and make to order strategy (pull system) for SKUs and customers with high variability (less than 50% of sales volume).
- For high variability customers, POS information is used to understand consumer profile and trigger replenishment process.
- Planners work closely with customers' demand planners to align weekly and monthly promotion calendar and market activities.
- Performance of planners is tied to compensation and rewards.
- Senior management understands, support and value demand planning function and use it to drive business results.
- Both statistical forecast tool and demand visibility in the supply chain are in place to generate forecast and define replenishment needs.
- Standard multidimensional metrics are used and cover forecast accuracy, customer service, inventory levels and costs, and are aligned with different echelons (e.g., customers) to share performance goals in the supply chain.

Level 4

- Demand planners have all required information to sense actual demand and use it as an input to the planning process.

- More than 60% and less than 80% of the company sales volume is sold using a pull system.
- Management has a clear focus and goal to reduce demand variability due to end of the month loading process, price discount to high volume customers or special consumer promotions (actual performance shows less than 40% variation between high and low peak weeks during the month).
- Sales target quota is set in a way to reduce demand variability and stimulate sell out volume (not sell in to other supply chain echelons).
- Organizational structure aims to synchronize all internal functional areas to become integrated and agile, providing a high customer service level (perfect order greater than 95%) on a monthly basis.
- Cross functional teams are developed to reduce silos and barriers between functional areas.
- Senior management creates a company culture centered in customer service and satisfaction, and market driven.
- Order process is based on service level agreements for demand driven replenishment. DD replenishment builds on the principles of lean manufacturing and pull based replenishment.
- Actual demand signals and forecasts are shared with suppliers and logistics providers on a daily basis. Alerts and exceptions are used to trigger supplier response to critical demand variations.
- Metrics:
  - Perfect order greater than 95% and profitability increase.

Level 5

- Same as in level 4, but in addition more than 80% of the company sales volume is sold using a pull system and only 20% remains using statistical forecast (mainly low variability SKUs).
- When used, statistical forecast shows high accuracy performance (greater than 90% on a SKU, week and plant level).
- Customer service mindset is full embedded in the company's culture and values.
- Marketing and Sales areas have a clear understanding about customers' needs and work closely with them to develop products and services that fit their needs.
- All new hired employees go through an orientation program in the first month to understand company's values, culture and behaviors as a demand driven organization.
- Actual demand information is available to all other functional areas (production, logistics, finance, procurement, etc.) through an IT tool and is used to make operational decisions that drives clear and proven bottom line results (reduction of inventory levels, unnecessary price reduction, reduction of product write off, etc.).

- Metrics:
  - Perfect order greater than 99%
  - Profitability increase
  - Supply chain cost as a percentage of net revenue shows clear downward trend in the last 3 years

5.2.2 Component: Demand Management

5.2.2.1 Category: Sales and Operations Planning (S&OP)

Level 1

- No formal S&OP process is in place.
- There are sporadic and informal meetings without formal agenda and participants.
- No integrated tool to support S&OP process.
- S&OP does not show clear bottom line results and improvements.
- Senior management does not understand, support or value S&OP as a management tool towards supply chain operational excellence.
- No standard or multifunctional metrics are revised and discussed during the meetings.

Level 2

- There is a formal monthly Sales & Operations Planning (S&OP) process that covers: (1) Data gathering, (2) Unconstraint statistical forecast, (3) Demand Planning, (4) S&OP analysis, (5) Pre-S&OP meeting, (6) Executive S&OP meeting.
- There is a formal weekly Sales & Operations Execution (S&OE) meeting to review operational plans against actual performance, and manage demand and supply variability.
- All functional areas actively and regularly participate in the meeting (finance, logistics, manufacturing, sales, marketing, procurement, IT, quality, etc.).
- Supply and demand plans are reconciled to generate one integrated operational plan, but the focus is still on pushing volume in the end of the month.
- Senior management supports and value S&OP process.
- S&OP tool interfaces with supply and demand systems to collect data and present both plans and KPIs results.
- Standard metrics are reviewed, discussed and cover Order Fill Rate, Forecast Accuracy, Inventory Turns (volume and dollar), Functional costs vs. budget.

Level 3

- Same as in level 2, but in addition monthly meetings discuss and revise separate plans for push and pull volumes and integrate them into one final plan.
- Demand signals are shared and aligned during the meeting across all functional areas.
- Standard metrics are reviewed, discussed and cover customer service, forecast accuracy (for push volume), demand error (for pull volume), working capital, total supply chain costs.

Level 4

- S&OP is embedded in the organizational culture and processes and is seen as a value added process towards high customer service and lower supply chain costs by all functional areas.
- Beyond monthly meetings, there are also ad-hoc meetings whenever there is a demand signal captured in the market that deserves attention or requires specific action plan.
- S&OP team has a proven track record of inventory reduction, customer service increase and higher profitability.
- Ownership of meetings is held by all functional areas.
- There is some external collaboration both with customers and suppliers to bring extra market information and demand alignment to the supply chain.
- Analyze lift for demand shaping, includes promotion planning, price management, and contract compliance with key customers. Evaluate the "what if" demand shaping based on profitability, revenue, customer service, and working capital.
- Identify constraints, and demand shortfalls.

Level 5

- Demand plans are aligned with customers and supply plans are aligned with key suppliers to ensure product availability and reduce variability through information sharing.
- An advanced S&OP tool integrates information of company demand and supply plans with both customers and suppliers IT systems.
- Event driven meetings are scheduled whenever there is a change in the plan or a supply–demand imbalance or specific market opportunities.
- Senior management understands and uses S&OP as a key weapon to drive alignment in the supply chain with customers and suppliers.
- The timing and rollout plans for new product introductions are an active process inside S&OP meetings.

5.2.3 *Component: Demand Management*

5.2.3.1 Category: Collaborative Planning and Forecast Replenishment (CPFR)

Level 1

- No CPFR is in place with customers or suppliers.
- There are basic and completely independent transactional systems for the company and its customers and suppliers.
- Limited or no access to demand data from customers, and company does not provide demand data to its suppliers.
- Demand signals are not considered in the forecast process, and forecasts are not communicated to other supply chain partners.

Level 2

- CPFR is piloted only in a very limited number of key customers (e.g., larger national accounts) or critical suppliers (e.g., top five suppliers), but it is not considered a key strategy for improving customer service and reducing supply chain costs.
- There are limited internal customer POS data available from key customers through EDI (Electronic Data Interchange), but the information is not formally integrated with the demand planning process. (POS information is only used for other activities like category management).

Level 3

- CPFR is implemented in different customers and suppliers that use a pull system, and there is a formal written business agreement between company and each of the trade partners to collaborate in a CPFR effort.
- Retailer POS data, causal information and planned events are used to create a sales forecast that support the joint business plan. The exception items that fall outside the threshold jointly defined by manufacturer and customer are discussed and resolved through email or formal meetings defined in the business agreement.
- An integrated decision support system exist and provide customer and market data between company and its trade partners (e.g., customers and suppliers).
- The demand signal is based on warehouse withdrawals and inventory levels for manufacturing companies and on POS information for customers.
- Results show clear reduction of cost and demand variability, as well as improved service.

Level 4

- CPFR provides full visibility of customer forecasts, warehouse withdrawals and inventory levels, and POS data is used to adjust short term demand signal.
- Joint sales and orders forecasts are created and internally integrated into production and replenishment planning systems.

Level 5

- A collaborative, integrated and automated communication process exists, and actual POS customer data is used to generate the short-term demand signal and long-term forecast.
- Advanced technologies (e.g., web based application) provide information seamlessly between trading partners to enable collaborative planning, evaluation, and execution.
- Full access to and use of individual trading partner information (e.g., POS data and shopper research) drives joint business planning and measures.

5.2.4 Component: Demand Management

5.2.4.1 Category: Vendor Managed Inventory (VMI)

Level 1

- No VMI is in place with customers or suppliers.
- There is no demand and inventory visibility across different stakeholders in the supply chain.

Level 2

- VMI is piloted only in a few customers or suppliers, but it is not considered as a key strategy for improving demand visibility and reducing inventory carrying costs.

Level 3

- VMI is implemented and fully operational in customers with pull system like MEPs (Market Execution Partners), Authorized distributors, etc.
- There is a VMI tool that receives the daily demand and inventory volume by SKU, and based on internal replenishment algorithm, proposes the right volume quantities to replenish for each SKU location.
- VMI implementation shows clear reduction of inventory levels, smooth sales volume and better integration between company and customers or suppliers.

Level 4

- Same as in level 3, but in addition VMI is implemented in all customers and suppliers that use pull system.
- VMI parameters are aligned and regularly reviewed by company and the supply chain partner to ensure suggested order meets business requirements.
- Master data (inventory and sales volume) is regularly updated and well maintained.
- Relationship between partners is well established and there is a clear understanding about VMI importance, operational model implemented and benefits generated.
- Trust between supply chain partners is clearly in place, reducing the requirement for order confirmation by customers.

Level 5

- Same as in level 4, but in addition VMI tool is integrated with production planning tool and demand planning tool.
- Replenishment volume proposed by VMI automatically feeds the production planning process, reducing emergency production orders and unnecessary changeovers.
- There is proven operational improvement in other functional areas like production, transportation and distribution out of the VMI implementation.

5.2.5 Component: Supply and Operations Management

5.2.5.1 Category: Procurement

Level 1

- There is no formal Procurement area in place or only a Purchase department with a pure transactional focus.
- Lack of procurement skills and capability inside the organization.

Level 2

- There is a Procurement area who works very close and aligned with all functional areas (e.g., Supply chain, Marketing, Sales, etc.) to provide solutions to the business.
- Top critical suppliers are identified, mapped and carefully managed in terms of annual and monthly operational capacity, quality, cost and financial healthy.

- Service level agreements (SLAs) with critical suppliers are in place and define operational goals, penalties, contingency plans, etc.

Level 3

- Procurement has a strategic role to develop and manage the supplier base to meet company's short and long term objectives in terms of capacity, quality and cost.
- Forecasted volumes are shared and aligned with suppliers with monthly meetings, where it is also reviewed supplier scorecard and discussed/aligned action plans to be implemented to improve performance.
- Volume from Pull ("make to order") system is also discussed and aligned with suppliers to meet right delivery lead time and quality.
- Critical Procurement skills and competencies like SRM (Supplier Relationship Management), Negotiation, Communication, Collaboration is in place.

Level 4

- As part of the SRM (Supplier Relationship Management) company and top strategic suppliers work together to fulfill actual demand through a true demand driven process.
- Companies collaborate on having an aligned business plan, where they agreed on common objectives and goals, like:
  - Reduce order lead times of critical ingredients and/or raw materials
  - Share actual demand information with suppliers to reduce volatility and variability
  - Suppliers share inventory "on hand" information and also production master plan with the company
  - Implement a lean process, reducing waste and inefficiencies in the purchase/production process for both companies
  - Use supplier expertise to support innovation and new product introduction
- There is low finish goods and raw material write off in the supplier operation and there is a clear inventory reduction without lost of sales.

Level 5

- Same as in level 4, but in addition IT systems from both company and suppliers are integrated to share accurate and seamless information.
- Top suppliers have integration and collaboration with its own suppliers, optimizing the whole supplier chain.

5.2.6 Component: Supply and Operations Management

5.2.6.1 Category: Manufacturing

Level 1

- There is no formal process for continuous improvement or formal Root Cause Analysis (RCA).
- Some asset assessments have occurred, but no action plans have been developed/ implemented from the studies.
- Some informal communication is maintained, sharing out performance information, but organized review meetings within teams are not regularly conducted.
- Quality inspections and maintenance have started to be performed by operators.
- Some housekeeping initiatives are in place, but there are still some areas below minimum standards.

Level 2

- There are shift and function handover reviews in place and Plant and Line Performance KPIs are visually displayed to all associates. Line performance reviews occur at a minimum two times per week.
- Discussions center on key issues and there is a formal improvement process in place with input from associates. Communication expands to cover Quality, Safety and Environment.
- KAIZEN, which refers to the philosophy or practices that focus upon continuous improvement of processes in manufacturing, engineering, supporting business processes, and RCA are the formal processes for continuous improvement and failure analysis. A formal team is in place to manage this process. All events are documented and available for review (minimum of 25% of the plant population has participated in a KAIZEN event).
- 5S initiatives have been piloted in the operations, and time lines exist to expand to the entire facility.
- Execution is occurring on critical asset utilization standard operating procedures including setup reduction efforts, Start up, and Shut down, BEC (Basic Equipment Care), Lubrication, etc., for all lines and an extensive amount of support equipment.
- SLE (Unconstraint line efficiency) is analyzed daily, weekly and monthly to determine critical issues; action plans are developed and implemented. Unconstrained SLE is at or above 55% (plant).

Level 3

- Same as in level 2, but in addition there is a shift from a pure manufacturing Push Process to a hybrid Push–Pull system.

- Products are categorized into a push or a pull manufacturing strategy (make to stock or make to order, respectively) based on demand variability and production efficiency.
- Fifty percent of the plant has participated of a KAIZEN event. There is a formal process to share KAIZEN learning with other plants.
- 5S is fully implemented reducing safety risks and providing better operation conditions for workers.
- Line operators are responsible for all troubleshooting, maintenance and quality checks in the production lines.
- There is a production planning and scheduling optimization tool in place to minimize changeover time, increasing production throughput.
- Make to order products are produced in an efficient way, and shows clear reduction of inventory levels and product/raw material write offs over time.
- SLE (Unconstraint line efficiency) shows clear upward trend to or above of 65%.

Level 4

- Same as in level 3, but in addition Lean manufacturing is the key strategic foundation to become an agile production process. Organization has implemented and captured real benefits from the following efforts for example:
  - **Setup reduction achieved through the SMED approach** (Single Minute Exchange of Die)**:** There are advanced preparation of equipment needed for production allowing for a fast changeover from one operation to another. (This is the foundation to move from a batch mentality).
  - **Cells with improved layouts:** Cells are implemented as a building block to become an agile and high-speed supply chain. Machines of different types required to produce a product are located together. The focus shifts from the production to the product. Cells enable products to be produced in small lot or single-lot quantities.
  - **Front line supervisor leadership:** Supervisors are empowered to lead, coach and provide feedback for their teams towards a continuous improvement performance through KPI management, shift log meeting, weekly trend analysis and monthly meeting to review performance.
  - **Flexible workers and Multifunctional teams:** Workers can operate several types of machines and are cross-trained in different operations in the production process. Multifunctional workers also enable the operation to be executed with fewer employees.
  - **Kanban systems to pull product through the plant and the supply chain:** This system creates links between operations, notifying upstream operations when to move and make production nits.
- Manufacturing is a key enabler to fulfill customer requirements towards product customization and innovation, keeping at the same time high operational efficiency and low production cost.

Level 5

- Same as in level 4, but in addition Postponement is in place and is incorporated into the manufacturing strategic foundation to increase flexibility and agility and allowing producing increased number of innovative SKUs that will fulfill market demand.
- Postponement relies on modular design, common components, quickly deployment of manufacturing resources and cost-effective customization that occurs as late in the production cycle as possible.

5.2.7 *Component: Supply and Operations Management*

5.2.7.1 Category: Logistics (Warehouse and Distribution)

Level 1

Warehouse

- Warehouse is not oriented towards customer service. There is no service level KPIs (e.g., fill rate, perfect order) measurement in place based on original customer order.
- There is no automatic product priorization implemented in the order processing system or is not aligned with route to market strategy.
- Basic or no storage rack structures in place to maximize warehouse density (cases per square meter). Density is not prioritized to increase asset utilization.
- Warehouse layout is not formally reviewed on a regular basis to increase warehouse productivity and reduce safety risks.
- Single forklifts are used most of the time (greater than 50%) to perform product put-away, retrieval and truck loading.
- No 5S and basic housekeeping initiatives implemented.

Distribution

- Distribution operation is not oriented towards customer service (e.g., On time and In full delivery).
- There is no distribution fleet policy or it is not fully executed to keep fleet average age aligned with planned targets.
- Sales and delivery territories are not balanced through a strategic route planning tool.
- Routing optimization tool is implemented and shows reduction of number of trucks required, total Km driven and distribution cost per case.
- There is no track and trace solution to manage the distribution operation during the day.

Level 2

Warehouse

- Warehouse operation is well executed with good operational performance (e.g., increased total warehouse productivity, low safety incidents – less than 0.5% lost time incident rate – LTIR, good inventory accuracy – greater than 99%, and downward trend on total warehouse costs over time).
- Racking structures are implemented and increased warehouse density (e.g., more than 10%), maximizing asset utilization.
- Double forklifts are used most of the time (greater than 50% of the time) to perform product put-away, retrieval and truck loading.
- There is a performance management process implemented that increased productivity over 20% and developed front line supervisor leadership to lead, coach and feedback the warehouse team towards a continuous improvement process. There are shift log meetings performed in the beginning of each shift to discuss safety, KPI performance and allow two way communications with all warehouse employees.
- Full housekeeping in place, but no 5S fully implemented.

Distribution

- Distribution operation is well executed with daily dynamic dispatching based on customer orders received.
- There is a performance management process implemented that developed front line supervisor leadership to lead, coach and feedback delivery teams towards a continuous improvement process. There are weekly meetings performed to discuss safety, KPI performance and allow two way communications with all delivery employees.
- There is a fleet renew policy in place that defines best mix of truck types, right number of trucks and target average age. Fleet policy is executed as planned on a yearly basis.
- There is no track and trace solution to manage the distribution operation during the day.
- There is a basic return management process in place to track percentage of volume returned by reason code. Return levels are greater than 2% and less than 3.5%.

Level 3

Warehouse

- Same as in level 2, but in addition 5S is implemented in a sustainable way and it is fully owned by warehouse employees (e.g., supervisors, forklift operators, pickers, etc.).
- Warehouse employees are well trained with all skills required to excel in the job and deliver the expected customer service, keep good product integrity, high picking accuracy, low cycle time and low operational costs.

- Succession plan is tied to performance management, training curriculum on different logistics functions (e.g., distribution, transportation, etc.) and leadership capability.
- Warehouse operation provides flexibility and speed for pull customers to delivery expected orders on time and in full.

Distribution

- Same as in level 2, but in addition there is a manual tracking tool to compare actual vs. planned route both for Km and hours driven on a daily basis after the route.
- Information from the tool is used to identify variances and root causes and define action plans to adjust plan or execution, depending on the root cause.
- Return management is in place with a formal process to measure, track and correct all important reason codes. Both sales and distribution functions share responsibility for overall performance and returns are under control with less than 2% of volume dispatched.

Level 4

Warehouse

- Warehouse is fully oriented towards customer service, meeting expected perfect order goals (greater than 99%) based on original orders.
- Warehouse layout is designed to provide flexibility, high density and speed to cope with customer demand. Simulation tool is regularly used to review layout and labor requirements aligned with sales demand.
- Triples or Quad forklifts are used most of the time (greater than 50%) to perform product put-away, retrieval and truck loading.
- Warehouse is a key enabler to customization and innovation through skilled workforce, standard operating procedures and multifunctional teams.

Distribution

- Distribution operation is fully oriented towards customer service with right truck types that maximize legal load limit, and meet customer requirements.
- There is a proof of delivery signed by each customer after receiving the products, using a handheld or smart phone. Information is sent to the company's control center after delivery and is used to validate deliver operation.
- Real time track and trace tool is implemented and allows managing and acting to solve distribution problems during the delivery route. There is a closed loop process that feedback actual information to the route optimization planning tool to improve compliance to plan of the distribution route.
- Estimated time of arrival (ETA) is provided to all customers in the beginning of the day.

Level 5

Warehouse

- Same as in level 4, but in addition warehouse performance metrics are included in the customer service level agreement (SLA) signed with customers.
- There is a monthly report delivered to each customer, showing all transactions performed, final service level delivered by the company, main reasons for failure and action plan implemented to increase service.
- Product Out-of-Stock in the market (not OOS in the warehouse) is a key performance indicator in the warehouse metrics.

Distribution

- Same as in level 4, but in addition Estimated Time of Arrival (ETA) is dynamically updated as the route is executed during the day and is sent to customers to ensure right deliver.

5.2.8 Component: Supply and Operations Management

5.2.8.1 Category: Customer Service

Level 1

- Customer service is not a strategic priority and is not defined as a critical operational capability.
- There is no formalized customer service policy and there is no dedicated customer service structure in place.
- Customer segmentation is not performed or is only defined based on company's sales volume (internal view).
- There is no regular customer service market research like Customer Value Driver to identify which services customers value more or Every Dealer survey, which aims to identify the true customer base.
- There is little o none formal collaboration between company and customers.

Level 2

- A formal customer service policy exists, and customers have been segmented along specific groups based on sales volume, market channel or local market criteria, but no process is used to build a fact-driven, customer-focused solution.
- There is no dedicated customer service structure in place. Commercial area is perceived as the mainly function responsible to deliver the customer service goals.

- There is little or none formal collaboration between company and customers in multifunctional areas like Marketing, Supply Chain, etc.
- No relation exists between execution and compensation or metrics.

Level 3

- Customer service policy is in place and fully executed.
- Formal market research approaches like Customer Value Driver and Every Dealer Survey are performed on an annual basis to review and design the customer service policy.
- Customer service policy also differentiates when using a pull or push supply chain approaches.
- There is no formalized customer service structure in place, but there are formal people from different areas of Marketing, Sales, Supply Chain, etc. assigned to support the development of solutions to customer requests.
- Market execution of the service policy is measured and is tied to compensation for the sales department.
- There is a deep understanding of value chain economics and cost-to-serve by various segments and/or customers.

Level 4

- Company has a formal and robust framework for developing a customer service policy (CSP) that drives capability and flexibility to profitably access all points of interaction with shoppers/consumers.
- CSP exists and is a dynamic, live document that defines roles, responsibilities, services to be delivered to each business segment/channel, special projects and initiatives to be implemented (e.g., Electronic Data Interchange, CPFR, VMI, etc.).
- There is a formal and dedicated customer service organization structure that reports to the company's CEO and is responsible to be the single point of contact to each segment, develop and implement customized solutions with multifunctional areas, and solve daily operational problems.
- Process and analytic capability is in place to dynamically optimize distribution to improve efficiency and maintain appropriate effectiveness/service levels through different service options.

Level 5

- Same as in level 4, but in addition customer service structure is organized on cells that have representatives from different functional areas like Sales,

Marketing, Supply Chain, Finance, etc. and each cell is responsible to manage a single or a group of customer segments.

- This structure provides quick, robust and customized solutions to customers, generating high product availability in the market, high customer satisfaction measured based on annual formal research performed with each customer segment and high level of collaboration across different functional areas.

5.2.9 Component: Supply and Operations Management

5.2.9.1 Category: Senior Management Support

Level 1

- Senior management does not understand or support Supply and Operations Management.

Level 2

- Senior Management understands Push Supply and Operations Management and uses it to meet business plan goals.

Level 3

- Senior management clearly understands the difference between Push and Pull strategies and its impact into supply and operations management and performance. They provide strong support to implement a hybrid Push and Pull.

Level 4

- Senior management understands the benefits of becoming a demand driven supply and operations management and uses it to improve customer service, reduce supply chain costs and meet business plan goals.

Level 5

- Senior management understands and supports the implementation of Pull strategy across the supply chain, and also fosters the collaboration in supply and operations management from suppliers to customers.

5.2.10 Component: Product Lifecycle Management

5.2.10.1 Category: New Product Forecast Models

Level 1

- Forecast for new products is generated based on management target volume or basic models like market coverage or market share. This forecast is the input for raw material planning and financial planning.
- Limited amount of data available for analysis or to be used in the forecast process.
- Operational problems usually increase forecast error (e.g., raw material availability, Out-of-Stock, low inventory accuracy, etc.)
- Organization does not have capability to quickly adjust and react to demand signals when launching new products.
- Low forecast accuracy (e.g., less than 50% FA at SKU level) with high demand variability.
- High percentage of product write off (>5%) and Out-of-Stock in the market for new products (>5%).

Level 2

- Forecast for new products is generated taking into consideration both the demand that is just for Pipeline fill (loading of inventory into channel member distribution centers and retail store locations) and the demand that is for Consumer demand (true consumer demand after the product is stable in the market).
- The forecast is the input for raw material planning and financial planning.
- Forecast is generated consider either quantitative models (e.g., Trend analysis, Exponential Smooth technique and Looks Like analysis, etc.) or qualitative models (e.g., Sales Force Composite, Assumption Based Model, etc.).
- Organization is still not able to quickly react to demand signals and suffers from Out-of-Stock when launching new products.
- Forecast accuracy is between 50 and 60% at SKU level.
- Product writes off is still high (>3% and less than 5%) and Out-of-Stock is greater than 3%.

Level 3

- Forecast for new products is generated only for customers under the Push system. For the remaining part of the business, under the Pull system, the demand visibility allows the organization to sense demand signals and adjust based on actual demand and not on forecast.

- When forecast is used, planners apply the models described in level 2, but in addition, they can also combine both qualitative and quantitative models to improve accuracy.
- Forecast accuracy for new products is between 60 and 70% at SKU level and product write off is between 2 and 3% and OOS is in the range of 2–3%.

Level 4

- Forecast is generated primarily only for raw material planning due to lead time from suppliers.
- For finish goods, the organization is structured to sense the demand and produce based on the true market demand, dramatically reducing OOS in the market to less than 2% and also product write off to less than 2%.
- Demand information feedback Marketing and Product Development departments to make "real time" adjustments to fully capture market opportunities.

Level 5

- Demand signal is shared across the supply chain where the organization operates, providing actual demand information to all echelons that sense the demand and produce based on the true market demand, increasing speed and flexibility in the supply chain (e.g., manufacturers, suppliers, etc.).
- OOS and product write off metrics are a common measure across all the members of the supply chain and all of them share their performance and rewards.

5.2.11 Component: Product Lifecycle Management

5.2.11.1 Category: Supply Chain Approach for Innovative Products

Level 1

- There is no formal product classification to apply different SC approaches. (There is only an "A, B, C classification" based on volume or revenue).
- It is applied the same supply chain strategic approach for new products or regular products.
- The major focus in the supply chain is to increase operational efficiency in each functional process, but not in an integrated way (e.g., functional optimization instead of supply chain optimization).
- The operation shows inventory level unbalance, low customer service level and over reliance on price promotions to meet monthly sales targets.
- Total supply chain cost is not clear measured and tracked, only functional costs.

Level 2

- Same as in level 1, but in addition there is an attempt to increase overall supply chain efficiency through integration of different functional areas, in order to minimize total supply chain cost, instead of isolated functional area optimization.
- Supply chain cost is measured and tracked, but still does not reflect a consistent downward trend over time.

Level 3

- Products are classified to use a Push or a Pull system based on demand and/or supply variability or maturity in a product lifecycle curve.
- Product classification is reviewed on a regular basis (every 6 months) or whenever needed to ensure the right supply chain strategy is in place.
- Products under the Push system are produced using a "make to stock" or "make to forecast" strategy. On the other hand, products under the Pull system are produced using a "make to order" strategy.

Level 4

- Each product is categorized either as an innovative product (e.g., great variety, short life cycles, high potential growth, volatile demand, etc.) or as a functional (regular) product (e.g., stable and predictable demand, long life cycles, lower growth rate, product does not change much over time).
- There is a clear understanding of two key supply chain functions – physical (focus on efficiency – maximize performance and minimize cost) and market (focus on responsiveness – product availability).
- Supply chain strategy is defined aligning product categories and supply chain functions as follow: Physical (efficient process) for functional products and Market (responsiveness) for innovative products.

Level 5

- Same as in level 4, but in addition there are specific operational actions fully implemented for each product category:
- **Functional products:**
  - High forecast accuracy for existing products with Push system
  - Inventory is well managed with high accuracy, reducing working capital
  - Reduced setups and emergency changeovers, increasing manufacturing capacity

- **Innovative products:**
  - Supplier agility and flexibility to quickly adjust and respond to demand variability
  - Inventory buffers are placed in the first echelons closed to actual demand to cope with uncertainty
  - There is inventory visibility in the supply chain

5.2.12 Component: Product Lifecycle Management

5.2.12.1 Category: Risk Assessment and Management for New Product Introduction

Level 1

- There are no risk assessment process and tool in place.
- New products are launched, but there is no formal process to gather and document practical learning and experience to be applied in new launches.

Level 2

- No risk management process and tool in place. However, after the introduction of each new product, the project team formally meets and reviews the process, learning, failures and successes, and document and stores them in a knowledge repository to be considered when launching other new products.

Level 3

- There is a basic and informally risk assessment process that is applied for all new products launched under the Push strategy.
- Based on assessment results, project team defines actions to be implemented during the project to reduce critical risks identified during the assessment.

Level 4

- There is a formal risk assessment process in place that is performed before the introduction of each new product, in order to reduce the risk associated with sales fluctuation and/or operational processes capabilities.

- There are several different risk factors included in the assessment like:
  - Level of innovation, time to market, product shelf life, manufacturing complexity, raw material characteristics, distribution complexity, expected profitability, etc.
- Based on risk assessment results, project team identify the most critical potential risks and develop mitigating actions to reduce exposure.
- After the introduction of each new product, the project team formally meets and reviews the process, learning, failures and successes and documents them and stores in a knowledge repository to be considered when launching other new products.

Level 5

- Same as in level 4, but in addition when starting a new project, there is a formal review of knowledge and learning from previous similar product launched in the past to ensure that previous mistakes and problems do not happen again.
- FMEA (Failure Mode and Effect Analysis) is applied to systematically examine each important aspect of the product launch to identify, prioritize and set mitigation strategies for potential gaps/weaknesses in new product introduction.

5.2.13 Component: Product Lifecycle Management

5.2.13.1 Category: Product Tracking and Visibility

Level 1

- There is no formal process to track product performance in the early phases of product launch in the market.
- There is no formal process to manage and track product retirement (SKU rationalization).

Level 2

- There is a formal tracking process for all new products introduced in the market that evaluates several different measures like:
  - Total volume evolution
  - Volume mix by market channel
  - Repeat purchase frequency evolution
  - Repeat purchase frequency by market channel
  - Price compliance

- Internal sales volume information is also evaluated and compared against sales target for new products. Whenever there is a gap, marketing and sales area defines and executes an action plan to improve product performance.
- There is a basic process to track volume of products to be discontinued (retired) from the portfolio, but there are frequently write offs for raw materials and finish goods.

Level 3

- There are different processes to track new product introduction for Push and Pull customers:
  - Pull customers: Actual daily demand information received from Pull customers are used to track new product performance in the early phases of the launch.
  - Push customers: There is a comparison between planned parameters used to forecast demand like product coverage, price to retail and to consumers, market share, inventory turns, etc. vs. actual performance in the market.
- There is a formal process to track volume and inventory for all products to be retired. Exceptions are identified and there are clear actions implemented to reduce product or raw material write offs, which shows a downward trend in the last 2 years.

Level 4

- Demand driven tools like CPFR, VMI, and Inventory visibility provide actual demand and inventory information between customers and the manufacturing company increasing agility and flexibility to adjust production and distribution based on real demand signals.
- Product retirement strategy is communicated and aligned between customers and manufacturing company. Both companies share demand and inventory information for each product to be retired, clearly reducing product and raw materials write offs and need to large price discounts.

Level 5

- New product launched information is shared across the supply chain, starting from POS (customer sell out) information all way through the suppliers, reducing bullwhip effect and OOS in the market place.
- Regarding product retirement, inventory and demand information is shared across the supply chain to provide visibility to all echelons and reduce write offs in the different partners in the chain. (There is a clear trend of write offs reduction in all the echelons of the supply chain).

5.2.14 *Component: Product Lifecycle Management*

5.2.14.1 Category: Portfolio Optimization

Level 1

- There is no formal portfolio optimization process in place. Emphasis is only on new product introduction without formal review and identification of products to be retired. (Discontinued from the product portfolio).

Level 2

- There are some attempts to take out products with low performance (e.g., declined sales or market share volume, products without clear strategy, etc.) from the company's portfolio. However, this process is not integrated with all critical functional areas (e.g., Marketing, Sales, Supply Chain, and Finance) and is not effective in discontinuing low performance products from the portfolio.
- There is no clear increase in supply chain efficiency and better commercial execution in the market.

Level 3

- There is a basic portfolio optimization process in place and performed on a regular basis, but the main criteria to retire a product is only on sales and market performance.
- Information from Pull customers are also gathered and considered as input to understand product performance and its strategic role in the total portfolio.
- There is a proven number of SKUs retired from the portfolio based on the set of criteria defined inside the organization.

Level 4

- There is a formal portfolio optimization process in place and executed on a regular basis (e.g., 2 times a year) to evaluate product portfolio and identify SKUs with underperform sales, lack of clear market strategy or low margin contribution to the company, etc.
- All SKUs identified are evaluated through different areas like Marketing, Sales and Supply chain, and information from Pull customers is essential to understand fit of product to strategic role in the portfolio.
- A final list is sent to a portfolio committee group that considers senior management participants from all critical areas of Marketing, Sales, Supply Chain,

Strategic Planning, and Finance. The committee reviews the proposed list of SKUs to be retired and make decision of keeping or retiring each product.

- There is a formal and updated RACI matrix (Responsible, Accountable, Consulted and Informed) that states clear roles and responsibilities for each participant in the process.
- Portfolio optimization process is effective in keeping a healthy and balance product portfolio over time.

Level 5

- Same as in level 4, but in addition, there is a portfolio optimization tool that support the execution of the entire process, manage exceptions and project timeline for SKU retirement across the supply chain (from customers to suppliers).

5.2.15 Component: Product Lifecycle Management

5.2.15.1 Category: Senior Management Support & Organizational Culture

Level 1

- Senior management does not understand or support the Product Lifecycle Management Process.
- Organization culture does not foster an innovative environment to bring new products in a faster and successful way, and does not motivate to review the product portfolio.

Level 2

- Senior Management has limited understanding of Product Lifecycle Management Process, but does not actually support the full implementation of PLM concepts inside the organization.
- Organization culture still does not provide the open environment for innovation and creative thinking. Functional silos and departmentalization is clearly perceived.

Level 3

- Senior management clearly understands and provides strong support to implement a robust PLM process that covers both new product introduction and portfolio optimization approaches.

- Organization culture provides open environment for discussion and people are not afraid to propose and implement innovative ideas.

Level 4

- Senior management uses PLM process as part of the annual business planning cycle to ensure alignment of the product portfolio with the overall company business plan strategy.
- Organization culture clearly foster an innovative thinking supported by annual objectives, rewards and recognitions programs aligned with annual performance management process.

Level 5

- Senior management supports PLM process aligned across the entire supply chain (from suppliers to customers) and not only inside the organization.
- Suppliers and customers are formally involved in the process and provide critical inputs for new product introduction and product retirement.
- Organization culture foster a customer driven mindset throughout the company and a supply chain focus on working collaboratively with different suppliers and customers towards a total supply chain optimization solution.

Chapter 6
Analytic Hierarchy Process

This chapter reviews the Analytic Hierarchy Process (AHP) method and explains how this method is applied to generate the weights for each category on the DDSC assessment.

6.1 Introduction to Analytic Hierarchy Process

The analytic hierarchy process (AHP) was developed by Thomas L. Saaty (1980) and is the well-known and useful method to obtain weights of each alternative in a multiple criteria decision-making problem. AHP requires the decision maker to provide judgments about the relative importance of each criterion, and then, specify a preference for each decision alternative using each criterion. The output of AHP is a prioritized ranking of the decision alternatives based on the overall preferences expressed by the decision maker.

Xia and Wu (2007) state that AHP consists of three parts: The hierarchy structure, the matrix of pairwise comparison ratios, and finally, the method for calculating weights. In AHP, a decision maker is asked to estimate pairwise comparison ratios with respect to strength of preference between subjects of comparison, thus the AHP is deeply related to human judgment.

AHP has been used to support decision process for different problems like Assess supply chain risks when analyzing offshore sourcing alternatives for a US manufacturing company (Schoenherr et al. 2008), Assess and identify the best delivery network design method taking into account both qualitative and quantitative factors (Sharma et al. 2008), Select anti cancer drugs to be produced and distributed within the pharmacy department of a French hospital (Vidal et al. 2010), Perform value chain analysis of service and manufacturing activities of a global supply chain of a multinational construction equipment corporation (Rabelo et al. 2007), Develop a model to assess the performance of small to medium sized manufacturing enterprises (Norita et al. 2006), Perform supplier selection with multiple criteria in volume discount environments (Xia and Wu 2007), Model

P. Mendes, *Demand Driven Supply Chain*,
DOI 10.1007/978-3-642-19992-9_6, © Springer-Verlag Berlin Heidelberg 2011

location analysis of international consolidation terminals (Min 1994), Perform carrier selection (Bagchi 1989), Propose a customer oriented approach to the warehouse network evaluation and design using a combination of Analytic Hierarchy Process (AHP) and Mixed Integer Linear Programming (MILP) (Korpela and Lehmusvaara 1999).

AHP has also been applied together with case study methods in different areas like Evaluation of critical success factors of ISO 14001 implementation in a case study in Malaysia (Sambasivan and Fei 2008), Case study to the selection of a multimedia authoring system in software selection (Lai et al. 1999), Case study of design and evaluation of automated cellular manufacturing systems with simulation modeling and AHP approach (Chan and Abhary 1996), just to give some examples available in the literature.

It is not the goal of this book to provide a detailed description of how AHP works, therefore, anyone interested in a comprehensive review of how to use AHP method, please refer to Saaty (1980). For a literature review on the integrated analytic hierarchy process and its applications, please refer to the work developed by Ho (2008), where he reviews the five tools that are commonly combined with the AHP process, like mathematical programming, quality function deployment (QFD), meta-heuristics, SWOT analysis and data envelopment analysis (DEA).

6.2 AHP Applied to Demand Driven Supply Chain Assessment Model

In this book, it is proposed to apply the AHP method to identify the weights for each component of the Demand Driven Supply Chain, and also for each category within each component, in order to ensure a consistent comparison, and a reliable overall score performance in the demand driven supply chain maturity model.

The first step of the method consists of decomposing the problem into a hierarchy structure, as illustrated in Fig. 6.1.

The second step consists of calculating the weights for components and categories. To that end, the author proposes to apply the priority scale developed by Saaty (1980) which follows the structure in Table 6.1.

For the pairwise comparison, it will be applied the approximation method developed by Wolff (2008), that proposes to compare only one alternative with all others instead of making pair wise comparisons for all alternatives with each other, which considerably reduces the understanding of the method and the work required from decision makers in developing the comparisons, as well as facilitates the calculations.

This is a very important aspect for this work, as the methodology can be applied in different companies and countries worldwide, and the easier the approach, the better it will run. The only requirement of the method is to select the strongest component or category, and uses it as the basis for comparisons with all others, as it

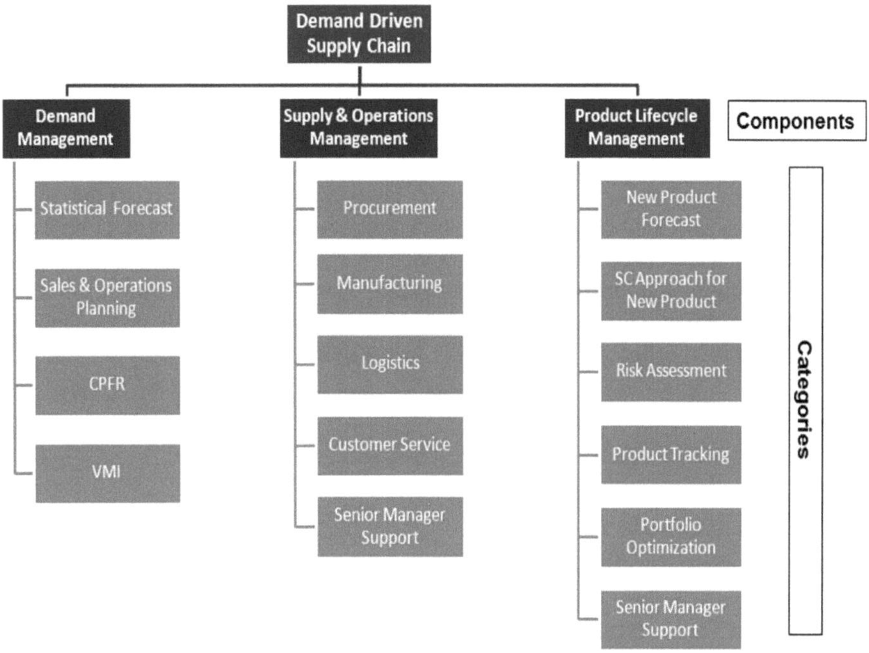

Fig. 6.1 Author's hierarchy structure for the demand driven supply chain model

Table 6.1 Priority scale for the pair wise comparison of components and categories in the DDSC model (Saaty 1980)

Priority scale to be used when defining weights		
Intensity of importance	Definition	Explanation
1	Equal importance	Two components/categories have the same importance
3	Slightly importance of one over another	Experience and judgment slightly favor one component/category over another
5	Strong importance	Experience and judgment strongly favor one component/category over another
7	Very strong importance	A component/category is favored very strongly over another
9	Extremely more important	A component/category is extremely more important over another
2, 4, 6, 8	Intermediate values	When compromise is needed adjacent scale values

will considerably reduce the probability of inconsistencies in the judgmental process. For more information about the approximation method, please refer to Wolff (2008).

For the practical case implementation, based on the author's experience of the CPG industry where the validation study was applied, it is considered the following components as the basis for comparisons:

- Component:
  - Supply and Operations Management should be the basis due to its large impact in both cost and customer service.
- Categories:
  - For Demand Management, Statistical Forecast should be the basis due to the industry still applies the make to stock approach to optimize its asset base and reduce fixed cost.
  - For Supply and Operations Management, Manufacturing should be the basis due to the importance of having low cost associated with producing the products in order to be competitive in the market place.

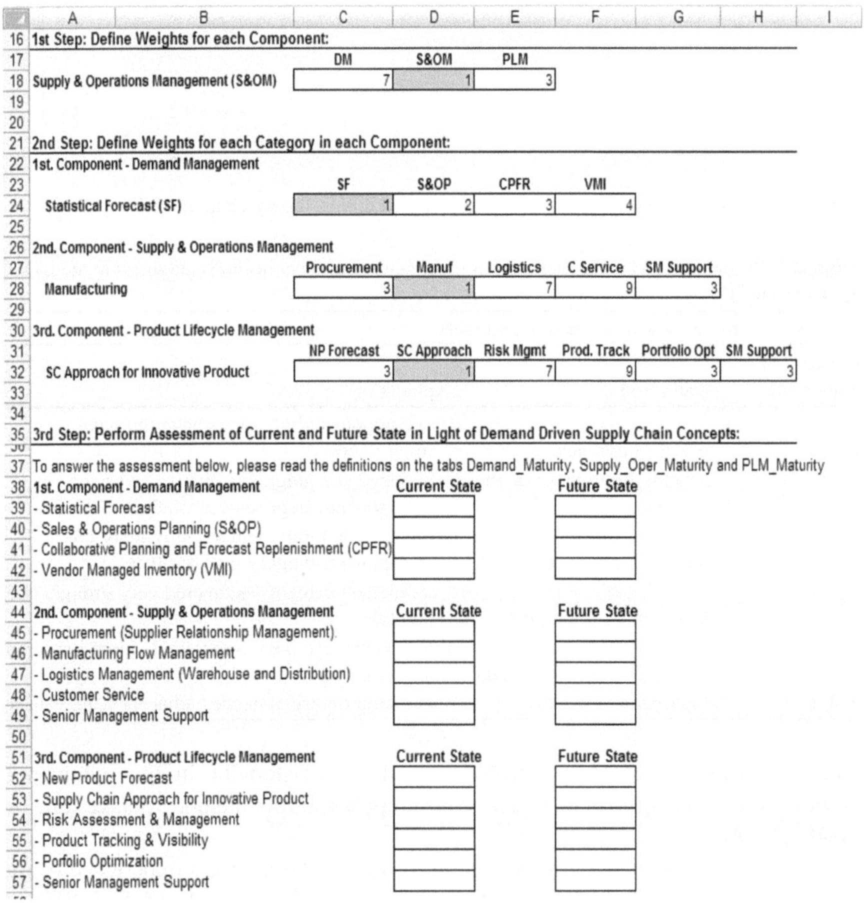

Fig. 6.2 Example of the author's spreadsheet to input answers

– For Product Lifecycle Management, Supply Chain Approach for Innovative Products should be the basis due to the strategy of launching new products in a fast pace to capture value and differentiate the company with customers and consumers.

The third step of the method consists of having supply chain directors identifying current and future states scores for their operations based on the definitions available in the maturity model which has 5 levels that were already described: Level 1 – Basic Push Operation, Level 2 – Optimized Push Operation, Level 3 – Hybrid Push – Pull Operation, Level 4 – Advanced Pull Operation, Level 5 – Optimized Pull Operation. For instance, if the SC director, after reading the maturity model, believes that his operation is still in level 1 for current state, he should enter number one into the respective field of current state in the spreadsheet.

It is important to highlight that it is not proposed in this book that all companies need to move to a high level 4 (advanced pull) or 5 (optimized pull) immediately, but instead, companies need to evaluate the competitiveness level of the marketplace where they operate, their organizational structure maturity, its supply chain complexity, which is aligned with the contingency theory described in the literature

	A	B	C	D	E	F	G
1	**Demand Driven Supply Chain - Summary Results for CURRENT STATE**						
2							
3		1. Demand Management		0,09		Current State	
4		Category	Weight within (1)	Overall Weight (2)	Score (3)	Score Within (1x3)	Overall Score (2x3)
5		Statistical Forecast	0,39	0,036	2	0,79	0,07
6		Sales & Operations Planning (S&OP)	0,08	0,007	2	0,16	0,01
7		Collaborative Planning (CPFR)	0,13	0,012	2	0,26	0,02
8		Vendor Managed Inventory (VMI)	0,39	0,036	2	0,79	0,07
9		Demand Management Score	1,00			2,00	0,18
10							
11							
12		2. Supply and Operations Management		0,45		Current State	
13		Category	Weight within (1)	Overall Weight (2)	Score (3)	Score Within (1x3)	Overall Score (2x3)
14		Procurement	0,12	0,054	2	0,24	0,11
15		Manufacturing	0,36	0,162	1	0,36	0,16
16		Logistics	0,12	0,054	2	0,24	0,11
17		Customer Service	0,05	0,023	2	0,10	0,05
18		Senior Management Support	0,36	0,162	2	0,71	0,32
19		Supply & Operations Management Score	1,00			1,64	0,75
20							
21							
22		3. Product Lifecycle Management		0,45		Current State	
23		Category	Weight within (1)	Overall Weight (2)	Score (3)	Score Within (1x3)	Overall Score (2x3)
24		New Product Forecast	0,05	0,022	1	0,05	0,02
25		Supply Chain Approach	0,33	0,151	1	0,33	0,15
26		Risk Assessment & Management	0,33	0,151	1	0,33	0,15
27		Product Tracking	0,07	0,030	1	0,07	0,03
28		Portfolio Optimization	0,11	0,050	2	0,22	0,10
29		Senior Management Support	0,11	0,050	2	0,22	0,10
30		Product Lifecycle Management	1,00			1,22	0,56
31							
32	Score	Description				Final Overall Score	1,48
33	1	Level 1 (Basic Push Operation)					
34	2	Level 2 (Optimized Push Oriented)					
35	3	Level 3 (Hybrid Push-Pull Oriented)					
36	4	Level 4 (Advanced Demand Driven - Pull)					
37	5	Level 5 (Optimized Demand Driven)					
38							

Fig. 6.3 Spreadsheet with example of current state scores

review done in Chap. 2, and then, identify the next level to move that makes sense both from a financial and operational perspectives.

For the case implementation, it was developed a standard spreadsheet that respondents just need to enter a few numbers and the spreadsheet provides the overall current and future states for 1 year horizon, based on the demand driven concepts, and also the individual scores for each category. The spreadsheet is illustrated in Figs. 6.2 and 6.3.

In the summary results spreadsheet showed in the example of Fig. 6.3, the supply chain director of the country under analysis can visualize the overall score result for his operation, as well as, the score results for each one of the three components of DDSC (e.g., Demand management, Supply and Operations management, and Product Lifecycle management).

In the specific example provided in Fig. 6.3, the overall score for the country is 1.48, which is in the middle of a basic (level 1) to an optimized push level (level 2), and each component has the following scores: Demand management (2.0), Supply & Operations management (1.64) and Product Lifecycle management (1.22).

In Sect. 7.1 will be presented the internet website developed by the author to automate the calculation presented in the spreadsheet 6.2 above, which is available for companies assess their current and future states on a regular basis.

References

Bagchi P (1989) Carrier selection: the analytic hierarchy process. Logist Transport Rev 25(1): 63–73

Chan F, Abhary K (1996) Design and evaluation of automated cellular manufacturing systems with simulation modeling and AHP approach: a case study. Integr Manuf Syst 7(6):39–52

Ho W (2008) Integrated analytic hierarchy process and its applications – a literature review. Eur J Oper Res 186(1):211–228

Korpela J, Lehmusvaara A (1999) A customer oriented approach to warehouse network evaluation and design. Int J Prod Econ 59(1–3):135–146

Lai V, Trueblood R, Wong B (1999) Software selection: a case study of the application of the analytical hierarchical process to the selection of a multimedia authoring system. Inform Manage 36(4):221–232

Min H (1994) Location analysis of international consolidation terminals using the analytic hierarchy process. J Bus Logist 15(2):25–44

Norita A, Berg D, Simons G (2006) The integration of analytical hierarchy process and data envelopment analysis in a multi-criteria decision-making problem. Int J Inform Technol Decis Making 5(2):263–276

Rabelo L, Eskandari H, Shaalan T, Helal M (2007) Value chain analysis using hybrid simulation and AHP. Int J Prod Econ 105:536–547

Saaty TL (1980) The analytic hierarchy process. McGraw-Hill, New York

Sambasivan M, Fei NY (2008) Evaluation of critical success factors of implementation of ISO 14001 using analytic hierarchy process: a case study from Malaysia. J Clean Prod 16(13): 1424–1433

Schoenherr T, Tummala V, Harrison T (2008) Assessing supply chain risks with the analytic hierarchy process: providing decision support for the off shoring decision by a US manufacturing company. J Purchasing Supply Manage 14:100–111

Sharma M, Moon I, Bae H (2008) Analytic hierarchy process to assess and optimize distribution network. Appl Math Comput 202(1):256–265

Vidal L, Sahin E, Martelli N, Berhoune M, Bonan B (2010) Applying AHP to select drugs to be produced by anticipation in a chemotherapy compounding unit. Expert Syst Appl 37(2): 1528–1534

Wolff C (2008) The analytic hierarchy process – conceptual review and proposal of simplification. M.Sc. Dissertation, Industrial Engineering Department, PUC/RJ

Xia W, Wu Z (2007) Suppliers selection with multiple criteria in volume discount environments. Omega 35(05):494–504

Chapter 7
Example of Case Implementation and Author's DDSC Website

This chapter explains how to use author's website to perform regular (e.g., annually) assessments in light of DDSC, and also provides practical examples of opportunities identified after applying the proposed methodology in a global CPG company. A framework is also presented to support the development of annual supply chain strategy to support company's progress towards DDSC concepts.

7.1 Step-by-Step Process to Perform Assessment Using Author's Website www.ddsconline.com

In order to make the assessment process easier and simple to be performed on a regular basis, the author developed all routines in his website. Figure 7.1 below provides the main page of the website, where users can login and access all assessment questions.

After login, users can visualize all previous assessments performed, as well as start a new assessment, as illustrated in Fig. 7.2.

When starting a new assessment, users should first rank the three components (Demand management, Supply and Operations management and Product Lifecycle management), and then, rank the categories within each component using the scale provided in the website. This is an important step as different industries should have different priorities for each component and each category. Figure 7.3 shows an example of the ranking alternatives.

The assessment process consists of 15 different questions, where user can score both current state and desired future state, as illustrated in Fig. 7.4 below.

When the assessment is completed, users can visualize their results in the main screen, as illustrated in Fig. 7.5.

DOI 10.1007/978-3-642-19992-9_7, © Springer-Verlag Berlin Heidelberg 2011

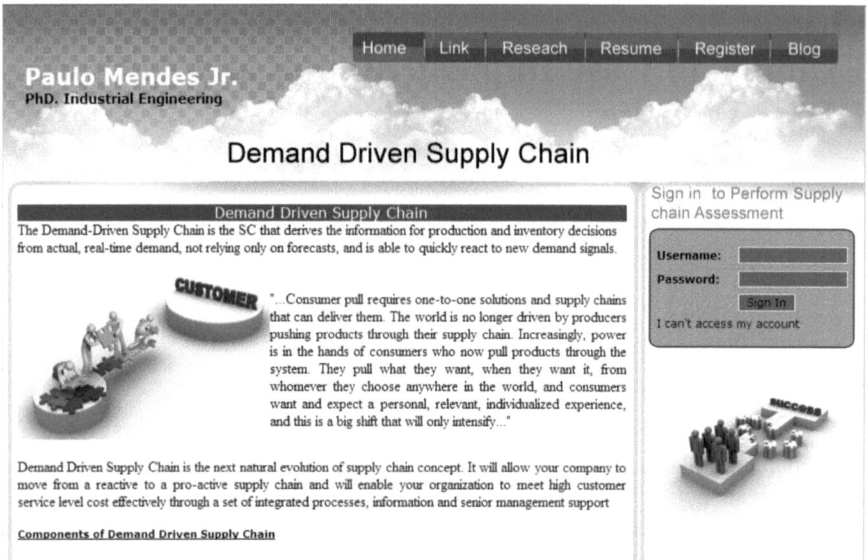

Fig. 7.1 Author's website to perform demand driven assessment

Fig. 7.2 Website stores all previous assessment performed

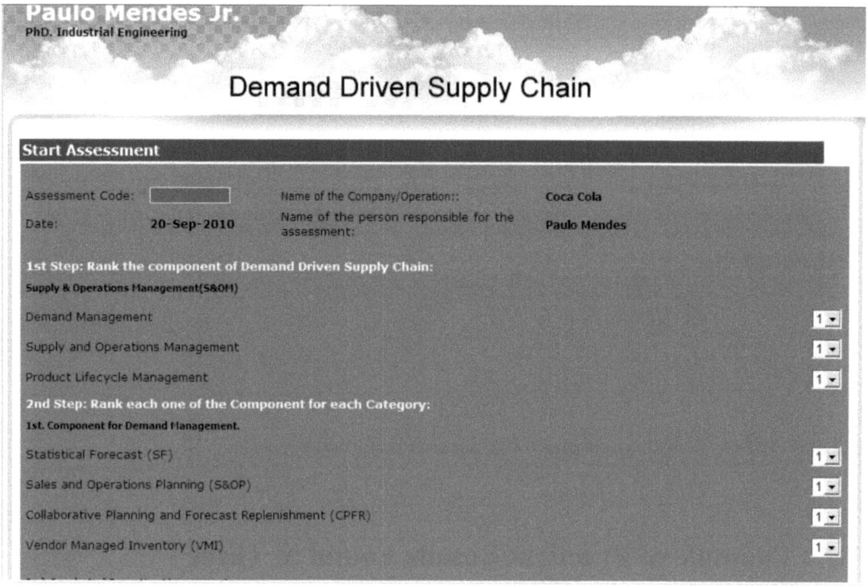

Fig. 7.3 Components and categories are ranked for each industry

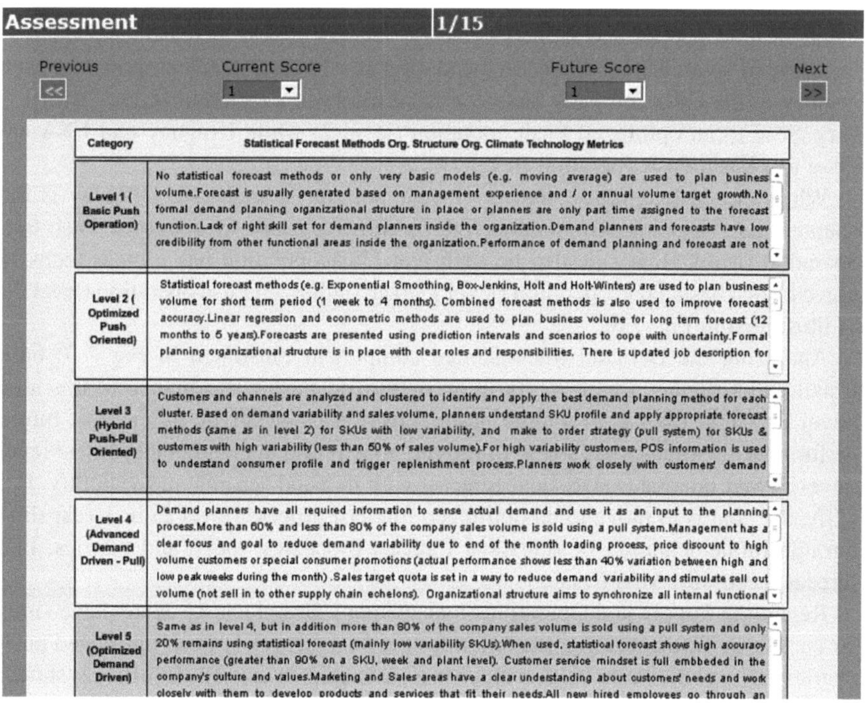

Fig. 7.4 Example of assessment question

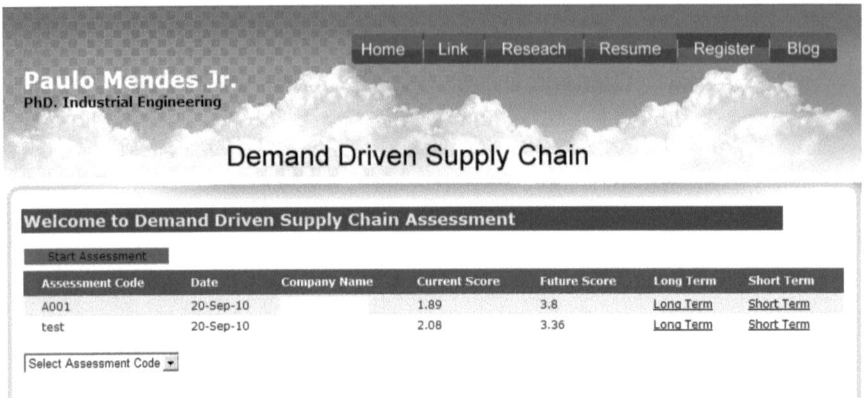

Fig. 7.5 Users can visualize scores for both current and future states

7.2 Example of Practical Results Found in Three Operations of a CPG Company

7.2.1 Overall Results

In terms of overall performance considering the three DDSC components, each country of the CPG company analyzed is currently in a different stage – Brazil is very close to an optimized Push operation (level 2) while Uruguay and USA are close to a basic Push operation (level 1).

When the desired future state in 1 year horizon is analyzed, Uruguay is the country that shows a strong objective to move closely to a hybrid Push-Pull operation (level 3). It can also be seen that USA operation has a clear focus to move from a basic Push operation (level 1) to an optimized Push operation (level 2), as illustrated in Fig. 7.6.

Analyzing the Demand Management component illustrated in Fig. 7.7, both Brazil and Uruguay are currently in an optimized Push performance in this area (level 2), while USA is in a more basic performance (level 1). In terms of future performance, Uruguay has a goal of moving to a hybrid demand management level (level 3) that integrates statistical forecast with demand sensing and visibility. For USA, the goal is to move to an optimized forecast process (level 2) in 1 year time through implementation of statistical forecast processes, tools, and metrics, like forecast accuracy, MAPE, etc.

Regarding Supply and Operations management, Brazil and Uruguay have similar current performance, in a transition from a basic (level 1) to an optimized push operation (level 2), while USA is close to a basic push operation (level 1), as shown in Fig. 7.8. For future performance, both Brazil and USA target to become an optimized Push operation (level 2), where senior management uses supply and

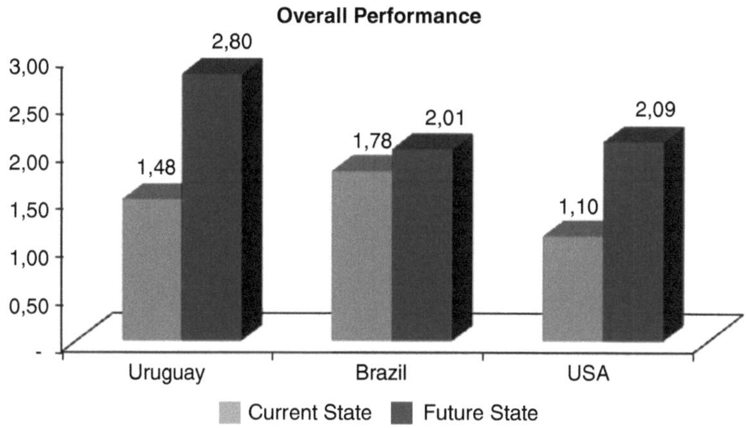

Fig. 7.6 Overall performance by country for current and future states

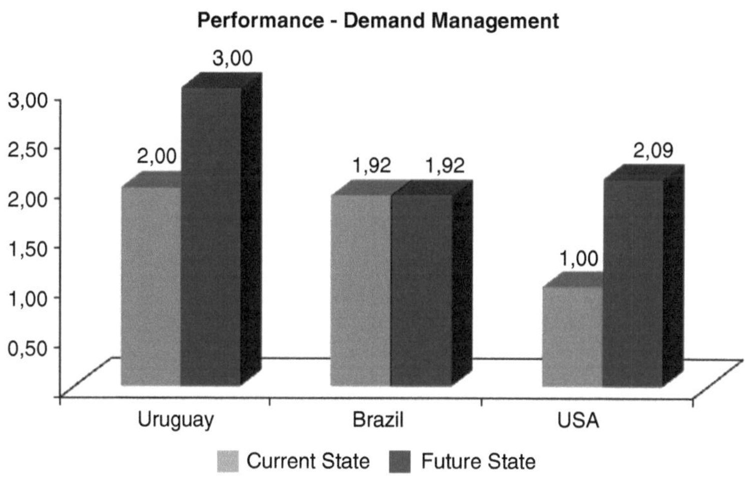

Fig. 7.7 Performance of demand management by country for current and future states

operations to meet business plan goals, while Uruguay target to move to a hybrid push-pull operation (level 3).

The last component of the DDSC is the Product Lifecycle Management (PLM). In this area, as it can be seen in Fig. 7.9, both Uruguay and USA currently have a basic push operation (level 1), which can be translated as a lack of an organization culture that foster innovation. Brazil is already in an optimized push performance (level 2) regarding PLM. For future performance, all operations target to become an optimized Push-Pull (level 3), where PLM is implemented for both new product introduction and portfolio optimization.

In the next section, the results for each country and the proposed action plan to support the improvement towards future state for each country will be detailed.

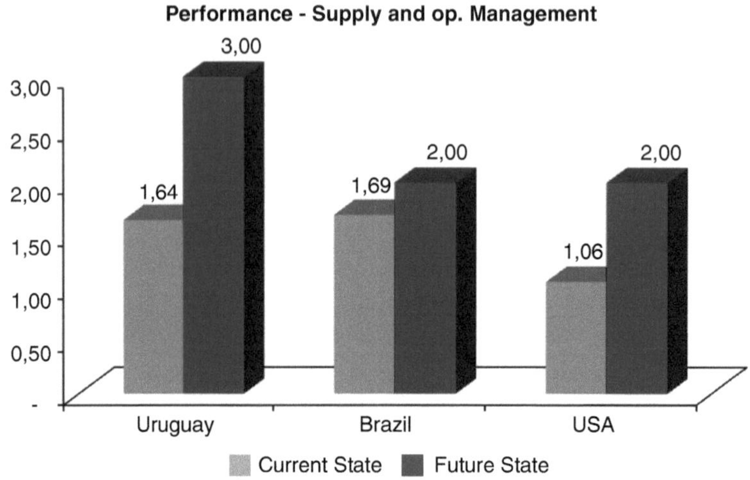

Fig. 7.8 Performance of supply and operations management by country for current and future states

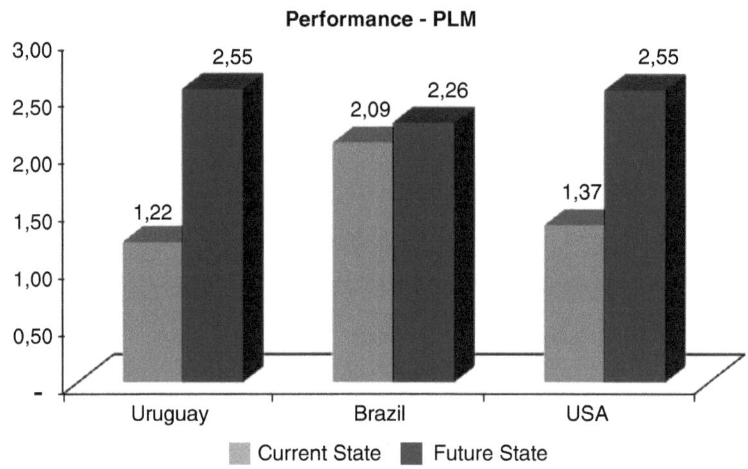

Fig. 7.9 Performance of PLM by country for current and future states

7.2.2 Example of Detailed Analysis for Two Operations Based on DDSC Assessment Results

7.2.2.1 Uruguay

Table 7.1 shows the results for both current and future states based on the answers provided by the leader of the supply chain area in Uruguay.

Table 7.1 Current and desired future performance for Uruguay

	Current state		Future state	
Category	Score	Weighted score	Score	Weighted score
Demand management				
Statistical forecast	2	0.79	3	1.18
Sales and operations planning (S&OP)	2	0.16	3	0.24
Collaborative planning (CPFR)	2	0.26	3	0.39
Vendor managed inventory (VMI)	2	0.79	3	1.18
Demand management score	–	2.00	–	3.00
Supply and operations management				
Procurement	2	0.24	3	0.36
Manufacturing	1	0.36	3	1.07
Logistics	2	0.24	3	0.36
Customer service	2	0.10	3	0.15
Senior management support	2	0.71	3	1.07
Supply and operations management score	–	1.64	–	3.00
Product lifecycle management				
New product forecast	1	0.05	2	0.09
Supply chain approach	1	0.33	2	0.66
Risk assessment and management	1	0.33	3	1.00
Product tracking	1	0.07	2	0.13
Portfolio optimization	2	0.22	3	0.33
Senior management support	2	0.22	3	0.33
Product lifecycle management score	–	1.22	–	2.55

In order to ensure the right focus and direction, the top two priorities were identified for each of the three DDSC components based on the highest weighted score, which, if well implemented and executed, will allow the organization to move quickly towards the desired future state.

For demand management, it is suggested to focus in **Statistical Forecast** and **Vendor Managed Inventory** as described below:

- For Statistical Forecast, it is important to define a process to formally analyze and cluster the SKUs sold in different customers and channels based on sales volume and demand variability, in order to apply an approach that combines statistical forecast for SKUs with low variability and actual POS demand information for SKUs with high variability. It is also suggested to implement a root cause analysis to map and understand the reasons of low forecast accuracy by SKU, and then, implement an effective action plan to fix the problems.
- For Vendor Managed inventory (VMI), the first step is to map current customer database to identify the group of potential customers to implement the VMI. Then, based on this list, commercial area together with supply chain should identify customers where actual POS information is readily available and could be shared on a daily basis with the company. Usually, potential customers for a successful VMI implementation are key account customers, like large retailers, airline companies, food chains, etc. One critical step in the VMI implementation

is to define and sign a memorandum of understanding (MOU) that specifies the operation of VMI process, Service Level Agreements (SLA), Penalties of noncompliance to defined goals, Backup process in case of IT problem, just to enumerate some of them.

For supply and operations management, it is suggested to focus on **Manufacturing** and **Senior Management Support** as detailed below:

- For Manufacturing, there are four key areas to focus:
  - Implementation of 5S or Workplace organization process to ensure a clean, safe and efficient workplace in each manufacturing line. This implementation is usually done through 5S events, where a "5S champion" will first train employees on each of the 5S phases (sorting, straightening, systematic cleaning, standardizing, and sustaining), and then, executes the changes at the end of the training of each phase.
  - Definition and implementation of a performance management process to develop supervisor leadership and employee engagement at the shop floor. This process should cover KPI definition and employee training, visual boards to post actual performance on a daily, weekly, and monthly level depending on the KPI, and handover meetings in the beginning of the shift to allow supervisors communicate the key focus and actions to be executed by line operators during the shift. Besides that, line operators should be responsible for all troubleshooting, maintenance and quality checks in the production lines.
  - Implementation of a formal process to perform "root cause analysis" of manufacturing SLE (system line efficiency) performance. This way, the company will be able to identify the main reasons that impact line efficiency and act to solve the problems.
  - Products should be categorized into a push or a pull manufacturing strategy (make to stock or make to order, respectively) based on demand variability and production efficiency.
- For Senior Management Support, the focus should be on providing senior management information about current performance and key actions to execute to allow meet the overall goal in supply and operations management. Usually, this could be done through development of an IT dashboard with critical KPIs from each functional area, and how these KPIs related and impact each other. Another focus is to educate senior management on the difference between Push and Pull strategies, and its impact into supply and operations management and performance.

For product lifecycle management, it is suggested to focus on **Supply Chain Approach** and **Risk Assessment and Management** as detailed below:

- For Supply Chain Approach, it is recommended to apply the proposed steps defined by Fisher (1997) to match type of product (functional or innovative) with the supply chain approach (efficient or responsive). For more information about this process, please refer to Sect. 4.5.2.2.

- For Risk Assessment, it is recommended to apply the proposed approach developed by the author in this book, where a formal risk assessment is performed before the introduction of each new product to allow developing mitigating strategies and ensure product availability in the market. For more information about the approach, please refer to Sect. 4.5.2.3.

7.2.2.2 Brazil

Table 7.2 shows the results for both current and future states based on the answers provided by the leader of the supply chain area in one of the regions inside Brazil.

In order to ensure the right focus and direction, the top two priorities for each of the three DDSC components were identified. However, for Brazil it is proposed to focus on the categories where the operation wants to improve in 1 year horizon, as the highest weighted score categories have the same current and future states.

For demand management, it is suggested to focus on **Statistical Forecast** and **Sales and Operations Planning (S&OP)** as described below:

Table 7.2 Current and desired future performance for Brazil

	Current state		Future state	
Category	Score	Weighted score	Score	Weighted score
Demand management				
Statistical forecast	2	0.79	2	0.79
Sales and operations planning (S&OP)	2	0.79	2	0.79
Collaborative planning (CPFR)	2	0.26	2	0.26
Vendor managed inventory (VMI)	1	0.08	1	0.08
Demand management score	–	1.92	–	1.92
Supply and operations management				
Procurement	2	0.46	2	0.46
Manufacturing	2	0.46	2	0.46
Logistics	1	0.23	2	0.46
Customer service	1	0.08	2	0.15
Senior management support	2	0.46	2	0.46
Supply and operations management score	–	1.69	–	2.00
Product lifecycle management				
New product forecast	1	0.09	2	0.18
Supply chain approach	2	0.53	2	0.53
Risk assessment and management	1	0.09	2	0.18
Product tracking	2	0.08	2	0.08
Portfolio optimization	2	0.53	2	0.53
Senior management support	3	0.79	3	0.79
Product lifecycle management score	–	2.09	–	2.26

- For Statistical Forecast, as the company already has a good forecast level with demand planning tools and processes in place, it is recommended to implement a root cause analysis to map and understand the reasons of low forecast accuracy by SKU, and then, implement an effectively action plan to fix the problems. Another opportunity is to implement medium to long term forecast tool to support the generation of forecasts for a longer horizon (e.g., 1–5 years).
- For Sales and Operations Planning (S&OP), the focus should be to stabilize the current process already implemented in the beginning of this year, to ensure it is engrained into the organization culture and generate sustainable improvement results. Another key area is to continue to look for senior management support on using S&OP as a key input for decision make process.

For supply and operations management, it is suggested to focus on **Logistics** and **Customer Service** as detailed below:

- For Logistics, it is recommended to focus on the following areas:
  - Implement 5S or Workplace organization process to ensure a clean, safe and efficient workplace in the warehouse. This implementation is usually done through 5S events, where a "5S champion" will first train employees on each of the 5S phases (sorting, straightening, systematic cleaning, standardizing, and sustaining), and then, executes the changes at the end of the training of each phase.
  - Analyze and re-design warehouse layout to increase density through the use of different racking structures. The second objective of the layout review is to increase warehouse throughput through the use of more efficient forklifts for put-away and retrieval operations.
  - For both warehouse and distribution, implement a performance management process to develop supervisor leadership and employee engagement. This process should cover KPI definition and employee training, visual boards to post actual performance on a daily, weekly, and monthly level depending on the KPI, and handover meetings in the beginning of the shift to allow supervisors communicate the key focus and actions to be executed in each warehouse area during the shift.

- For Customer Service, it is recommended to perform market research surveys like "customer value driver" and "every dealer survey" to understand customers' expectations and requirements, and to segment them accordingly. Based on the results of these researches, develop a formal written customer service policy which will define the service package that will be delivered to each segment/group of customers. This process should be reviewed on a regular basis (e.g., annually) to ensure that the company keeps its focus on the right priorities based on market needs. Another opportunity is to start the collaboration process with customers to align supply chain priorities in terms of product availability (e.g., fill rate), and operational efficiency goals for truck turnaround time, replenishment frequency, delivery time windows, etc.

For product lifecycle management, it is suggested to focus on **New Product Forecast** and **Risk Assessment and Management** as detailed below:

- For New Product Forecast, it is recommended to develop a formal process to understand in detail each new product introduction and define which forecast model (both quantitatively and qualitatively) is more appropriate for each new launch. Demand planning area should model forecast requirements for "pipeline fill" separately from "pure consumer demand", in order to be able to track and compare forecast vs. actual demand in the different phases of the new product introduction in the market. It is also necessary to develop an information database that will provide information about past launches performance, data required to apply the different forecast models, etc.
- For Risk Assessment, it is recommended to apply the proposed approach developed by the author in this book, where a formal risk assessment is performed before the introduction of each new product to allow developing mitigating strategies and ensure product availability in the market. For more information about the approach, please refer to Sect. 4.5.2.3.

7.3 Develop Supply Chain Strategy to Become Demand Driven

It is presented in this section, a proposed framework for supporting organizations on building a Demand Driven Supply Chain Strategy.

7.3.1 Introduction

The last step of the proposed methodology is to develop a supply chain strategy that will define all critical initiatives that the company should perform to move towards a demand driven supply chain. This is a very important step to ensure a structured and formal process to define and prioritize the different opportunities and strategic options available to the company.

Cohen and Roussel (2005) define strategic supply chain as a process to create a unique supply chain configuration that will drive the company's strategic objectives, and should consider five critical configuration components:

- **Operations strategy** refers to the decisions on how the company will produce its goods and services. Examples of decisions are which production strategy will be used (e.g., make to stock, make to order or some combination), what is the best balance between in source and outsource, will the company pursue a low cost offshore manufacturing strategy, just to enumerate some examples.
- **Outsourcing strategy** refers to defining what operational areas the company should keep in house and what areas should be outsourced. Areas where the company has expertise or that will provide a competitive advantaged for the

company should be kept in house. Outsourcing allows the company to ramp up or down quickly, build new products, or reposition itself in the marketplace by leveraging the expertise and capacity of other companies. This added flexibility and agility can make an enormous difference in today's competitive markets.

- **Channel strategy** refers on how the company will get its products and services to buyers and customers. These decisions address such issues as whether the company will sell indirectly through distributors or retailers or directly to customers, via the internet or a direct sales force. The market segments and geographies should drive decisions in this area, since profit margins vary depending on which channels are used and who gets the products in times of product shortages or high demand. Market leaders use effective channel strategies to reap significant gains like Dell, with its direct-sales model, and Wal-Mart with its superstore model, offer compelling examples of how channel choices can deliver a competitive advantage.
- **Customer service strategy** refers to prioritizing and focusing supply chain capabilities to deliver expected customer service.
- **Asset network** refers to the decisions on how the company will configure the supply chain to meet business goals. These decisions are usually related to factories, warehouses, production equipments, order desks, etc. The location, size and mission of these assets have a major impact on supply chain performance.

Cohen and Roussel (2005) also state that the configuration components – operations strategy, channel strategy, outsourcing strategy, customer service strategy, and asset network – are the fundamental building blocks of a supply chain strategy. However, to drive forward the company's strategic business objectives and really gain a competitive edge, these components should be aligned with the business strategy, with the customers' needs, and with the company's power and influence position relative to customers and suppliers.

Rodrigues et al. (2004) demonstrate the importance of aligning operational policies, procedures, guidelines, and training with high-level strategy and supporting it with appropriate information and measurement systems.

Morash, E. developed a strategy/capability/performance paradigm to link supply chain strategy with supply chain capability to ensure a sustainable high performance, as illustrated in Fig. 7.10 below.

Kaplan and Norton (2008) state that strategy develops and management is a closed-loop process where each part of the system influences all other parts. They proposed the framework shown in Fig. 7.11 to integrate strategy formulation and planning with operational execution.

As it can be seen, the strategy development starts by defining a high level vision of the organization's destination and finishes with executive leaders and teams launching the organization into action by implementing portfolios of aligned strategic initiatives. The framework in Fig. 7.11 has six major stages:

- Stage 1: Managers develop the strategy.
- Stage 2: Organization plans the strategy using tools, such as strategy maps and Balanced Scorecards.

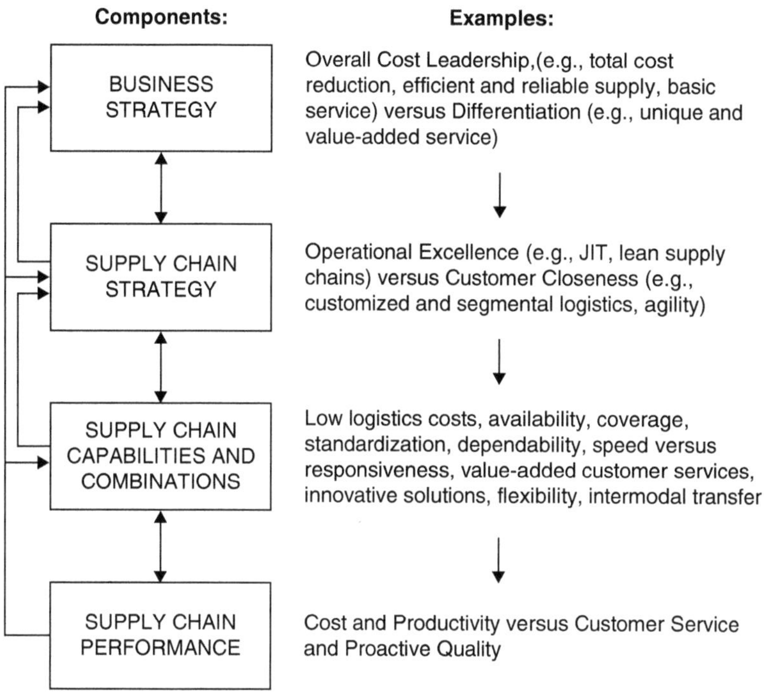

Fig. 7.10 Morash paradigm on supply chain strategy

- Stage 3: Once the high level strategy map and Balanced Scorecard have been defined, managers align the organization with the strategy by cascading linked strategy maps to all organizational units. They align employees through a formal communication process and link employees' personal objectives and incentives to strategic objectives.
- Stage 4: With all organizational units and employees aligned with the strategy, managers can plan operations using tools such as process management, reengineering, process dashboards, activity-based costing, resource capacity planning, and dynamic budgeting.
- Stage 5: As the strategy and operational plans are executed, the organization monitors and learns about problems, barriers, and challenges. This process should integrate information about operations and strategy in a carefully designed structure of management review meetings.
- Stage 6: Managers use internal operational data and new external environmental and competitive data to test and adapt the strategy, launching another loop around the integrated strategy planning and operational execution system.

In the next section, it will detail the author's proposed framework to allow companies developing a structured and formal supply chain strategy to become Demand Driven.

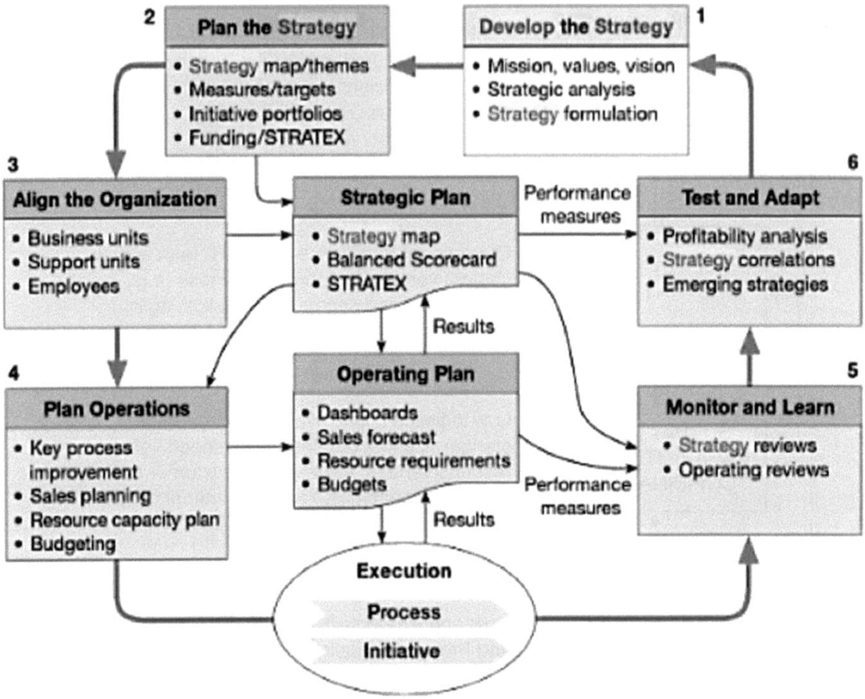

Fig. 7.11 Management system (Kaplan and Norton 2008)

7.3.2 Framework to Develop a Demand Driven Supply Chain Strategy

Figure 7.12 details the author's proposed strategic framework to develop a 3-year supply chain strategy to move towards a Demand Driven Supply Chain. The framework is divided in three major categories:

- **Inputs**: Refers to general market and business information that should be taken into consideration when developing a supply chain strategy.
- **Strategic Planning Development**: Refers to the planning steps required to develop the supply chain strategy.
- **Outputs**: Refers to the outcomes of the planning process and covers the SC strategic plan, organizational and capability requirements, and required resources and capital plans.

Each one of the components is described below.

7.3.2.1 Business Plan and Strategy

The first step to develop a SC strategic plan is to understand the company's strategic business plan for the next 3 years, as it will provide input and direction on the key

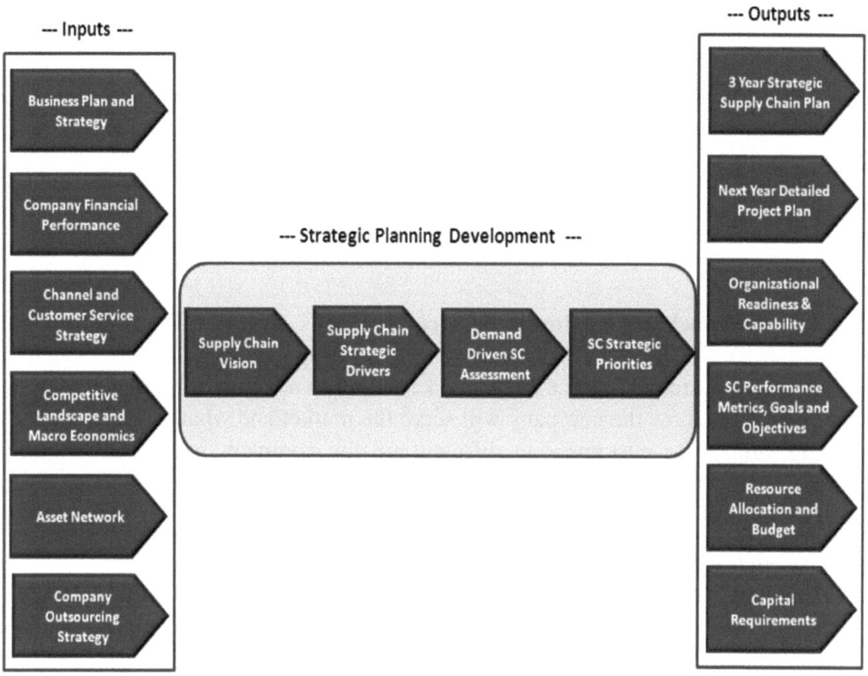

Fig. 7.12 Author's proposed framework to develop a 3-year DDSC strategy

strategies and initiatives that will be deployed in the market to deliver desired company's results.

Examples of business plan inputs:

- Company mission and vision.
- Company goals in terms of revenue and market share growth, finance performance in EBITDA and market value, customer service performance in customer satisfaction and order fill rate.
- Product and service innovation (new product introduction and SKU optimization).
- Organizational culture and values.
- People capability development.

7.3.2.2 Company Financial Performance

The second critical input to consider is the financial performance. It is necessary to understand both current financial performance and future goals, in order to link SC projects and initiatives with the business and financial objectives to be achieved.

Examples of financial metrics to consider:

- Cash Operating Cycle
- Return on Investment (ROI) and Return on Assets (ROA)

- Operating Income Margin (Profitability)
- Cash Flow requirements
- Fixed Asset utilization
- Days of Inventory (DOI)
- Total Supply Chain Expenses
- Manufacturing Cost per unit and also as a percentage of Net Revenue
- Logistics Cost per unit and also as a percentage of Net Revenue

7.3.2.3 Channel and Customer Service Strategy

Channel and customer service are also a key input to develop SC strategic plan, as they will define how the company will serve the market and what service packages will be offered to customers to differentiate the company in the marketplace. Supply chain should be a key enabler to deliver the expected customer service level.

Examples of inputs:

- Customer database growth by channel/segment
- Direct vs. indirect distribution (e.g., Wholesalers, Authorized distributors, etc.)
- Segmentation strategy (e.g., based on volume, market share, geography, channel, etc.)
- Market execution goals (e.g., share of inventory, priority product by cluster, etc.)
- Customer service (e.g., expected service packages, order fill rate goals, customized service requirements for sales and delivery, etc.)

7.3.2.4 Competitive Landscape and Macro Economics

To develop an effective DDSC strategy, it is critical to understand the competitive landscape where the company operates and the macro economic factors forecasted for the planning horizon. As the company moves into the future, it should be able to compare these forecasted factors against actual figures, in order to identify potential changes to the strategic plan.

Examples of macro economics inputs:

- GDP growth
- Interest rate
- Inflation rate
- Exchange rate
- Unemployment rate

Examples of inputs from competitive landscape:

- Competitor's execution in the market (e.g., operating culture, channel strategy, product strategy)
- Competitor's growth strategy and finance structure

- Competitor's supply chain structure (e.g., number of plants, DCs, manufacturing strategy, distribution strategy, procurement alliance, quality goals, etc.)

7.3.2.5 Asset Network

Understanding of current asset network capacity and operational capability is critical to identify potential gaps that could jeopardize sales growth and new product introduction.

Examples of asset network inputs:

- Location, capacity and number of manufacturing plants
- Location, capacity and number of warehouses
- Previous capital plans to expand new plants, production lines and warehouse facilities
- Changes in supply of key raw material (sugar, pre-form, label, caps, etc.)

7.3.2.6 Company Outsourcing Strategy

It is important to understand which operational areas the company should keep in house and which areas should be outsourced. Outsourcing can occur in different levels like individual activities, when outsourcing involves moving specific positions out of the organization, functional level, which involves moving some functions out of the organization, and process level, which involves outsourcing operational processes to other companies.

Examples of outsourcing questions to be considered:

- Current and future required core competencies
- Current and future organizational structure
- Current and future cost levels
- Current and future competitive advantages

7.3.2.7 Supply Chain Vision

After revising the critical inputs, supply chain directors should start developing the SC strategic plan. The first step is to define the supply chain vision. To that end, they should start brainstorming and defining a clear strategic vision statement that will be communicated and engrained throughout the organization, and will guide all initiatives.

Examples of points to consider when developing the supply chain vision:

- It should not reflect current performance, but instead, should inspire all associates in the SC organization to achieve a greater performance.
- It should consider different dimensions like cost, customer service, quality, safety, productivity.

- For critical dimensions, it should define a quantitative goal to be achieved (e.g., zero harm for employee safety).
- It should be developed together with different departments and job positions inside the supply chain organization to guarantee commitment and "buy in" from all personnel involved.

7.3.2.8 Supply Chain Strategic Priorities

After the vision statement, company should start brainstorming what are the key supply chain drivers that need to be developed and achieved, in order to generate the expected business results.

Examples of supply chain strategic drivers:

- Operational execution efficiency
- Employee safety performance
- Supply chain cost performance
- Supply chain agility
- Supply chain resiliency
- Demand driven focus
- Order fill rate goals (orders delivered on time and in full/total orders)
- Inventory management and working capital requirement
- Product quality performance
- Environmental responsibility

7.3.2.9 Demand Driven Supply Chain Assessment

The results and findings identified after performing the proposed DDSC assessment model provide relevant information on the key areas to focus in order to become a demand driven supply chain. The author's proposed assessment approach covers not only current state but also future state in 1 year horizon, which will allow identify the key areas to focus when developing the strategic plan.

Areas to be covered in the DDSC assessment:

- Demand management
- Supply and operations management
- Product lifecycle management

7.3.2.10 Supply Chain Strategic Priorities

The result of the strategic planning process should be a list of strategic priorities that will guide the design of a 3-year strategic plan. Every year, these priorities will be updated as the company performs new strategic planning cycle.

Examples of prioritized initiatives:

- Engrain safety in all operational processes to achieve "zero harm."
- Become a demand driven supply chain to eliminate inventory at the same time that delivers expected customer service level.
- Improve manufacturing efficiency and flexibility to cope with increase of SKUs in the product portfolio.
- Keep high quality product rating to differentiate our products in the market.
- Improve routing capability and control delivery execution to reduce miles traveled and ensure customer service level.

7.3.2.11 Three-Year Strategic Supply Chain Plan

Based on the strategic priorities and the financial impact, supply chain directors will be able to design the 3-year strategic plan which will define the initiatives to be deployed to achieve the desired Supply Chain vision.

Topics to be considered when developing strategic plan:

- It should cover all supply chain functional areas (e.g., procurement, manufacturing, logistics, safety, quality, engineering, etc.).
- Strategy design is the right moment to trade off conflict goals between different functional areas (e.g., manufacturing efficiency and inventory level, procurement lot size and inventory level, etc.).
- Set quantitative goals for each functional area, whenever possible, to make strategy tangible.
- Capability development is a critical enabler of the strategy, and therefore, should be detailed consider in the plan.

7.3.2.12 New Year Detailed Project Plan

After the prioritization discussion and alignment, company needs to generate a project plan on what will be accomplished in next year for each one of the opportunities identified during the strategic planning process, in order to ensure that each one of the strategic initiatives will be implemented as planned. It is also very important to assign clear responsibilities and timelines for each project.

Examples of points for a project plan:

- Project overview and expected cost
- Key activities to be performed in the project
- Estimated business impact (e.g., cost saving, cost avoidance, capital avoidance, etc.)
- Targets ("due dates")
- Capital required to implement the project
- Resources required

7.3.2.13 Organizational Readiness and Capability

Setting the right organizational structure, with right number of people and right set of skills in each position, with right compensation and performance management is a critical successful factor to implement the strategy and achieve the desired business results.

Examples of points to consider in the organization readiness:

- Job description updated and detailing key responsibilities and skills requirements based on supply chain strategy.
- Regular assessment of current "capability pipeline" inside the supply chain organization.
- People capability development plan based on gaps to execute the supply chain strategy.
- Performance management implemented and linked to compensation.

7.3.2.14 Supply Chain Performance, Metrics, Goals and Objectives

Integrating each different functional area (e.g., manufacturing and logistics) inside the supply chain department and also with all other areas inside the organization (e.g., commercial, finance) is a critical and difficult task, but extremely necessary. To achieve this objective, it is required to set cross-functional metrics that impact each functional area and forces the organization to work as a one integrated entity.

Examples of cross-function supply chain metrics:

- Manufacturing: Line efficiency, manufacturing cost per unit, days of inventory, inventory turns, etc.
- Logistics: warehouse utilization, days of inventory, inventory turns, and warehouse cost per unit, delivery cost per unit, delivery capacity utilization, etc.
- Quality: product quality index, production line efficiency, etc.
- Commercial: demand forecast accuracy, sales variability, inventory turns.
- Finance: supply chain cost per unit.

7.3.2.15 Resource Allocation and Budget

After validation of the key initiatives to be deployed, supply chain area should develop a detailed budget for the upcoming year, considering both the required resources to perform the operation and the expected savings that will be captured based on the initiatives planned.

Examples of points to consider in the budget process:

- Budget should be opened by functional area/department and by month.
- Compare new year vs. prior year budget to understand major variances by department.
- Cost should be aligned with forecasted sales volume.

7.3.2.16 Capital Requirements

It is also necessary to detail the required capital investment to capture the expected benefits in terms of efficiency or better customer service. Supply chain area should submit the capital requirements to the finance area, which will consolidate the overall company capital requirement and will submit it to the board of directors for approval.

Examples of capital requests in the supply chain area:

- Installation of new production lines
- Expansion of warehouse storage capacity
- Installation of new equipments to improve line efficiency or product quality
- Purchase of new trucks to keep up with sales
- Implementation of new IT tools to improve supply chain planning
- Supporting tools to implement demand driven processes like VMI, CPFR, etc.

References

Cohen S, Roussel J (2005) Strategic supply chain management: the five disciplines for top performance. McGraw Hill, New York

Fisher M (1997) What is the right supply chain for your product? Harv Bus Rev 75(2):105–116

Kaplan R, Norton D (2008) The execution premium: linking strategy to operations for competitive advantage. Harvard Business Press, Boston

Rodrigues A, Stank T, Lynch D (2004) Linking strategy, structure, process, and performance in integrated logistics. J Bus Logistics 25(2):65–94

Chapter 8
Summary and Future Developments

This chapter provides a summary of main contributions of this book to support companies move towards a demand driven supply chain operation, and also opportunities for future developments.

In this book, it is reviewed the current market and business environment and highlighted the importance of companies become demand driven instead of production driven based on the contingency approach. In a demand driven supply chain, customers activate the replenishment flow and the organization is structured and prepared to sense and respond to real time demand across the supply chain, which should include customers and suppliers.

The financial and operational benefits that emerge from applying the demand driven concepts were also reviewed, like reduced inventory levels, increased production efficiencies, decreased freight cost, and improved fill rates and product availability in the market.

In this book, a framework was also developed to allow companies to assess their current state in light of demand driven supply chain concepts, and identify the desired future state in a 1 year horizon. The framework was applied to three supply chain operations of a CPG company in different countries as part of the methodology evaluation. Results indicated that one operation is currently close to an optimized push operation (level 2 out of 5), and the other two operations are close to a basic push operation (level 1 out of 5), revealing that there are clear opportunities to implement the demand driven supply chain concepts to move towards a more customer centric operation.

This research also developed a formal framework to support companies in designing a supply chain strategy that will allow them to move towards a demand driven supply chain. This framework is integrated with the assessment process of both current and future states proposed in this thesis, and should be used as part of the annual planning cycle.

Finally, this book contributed towards a better understanding of the demand driven supply chain concepts and how to effectively implement those concepts in a real supply chain environment.

P. Mendes, *Demand Driven Supply Chain*,

DOI 10.1007/978-3-642-19992-9_8,

In terms of future developments, additional research could be done to cover the following points:

- Review maturity model for companies that do not have a manufacturing process, like service and retail companies.
- Measure actual benefits captured by companies after 1–3 years of implementation of the proposed methodology.
- Continue to further develop and refine the proposed framework to design a Demand Driven Supply Chain Strategy.

Index

P. Mendes, *Demand Driven Supply Chain*,
DOI 10.1007/978-3-642-19992-9, © Springer-Verlag Berlin Heidelberg 2011

MIX
Papier aus verantwortungsvollen Quellen
Paper from responsible sources
FSC® C105338
FSC
www.fsc.org

If you have any concerns about our products,
you can contact us on
ProductSafety@springernature.com

In case Publisher is established outside the EU,
the EU authorized representative is:
Springer Nature Customer Service Center GmbH
Europaplatz 3, 69115 Heidelberg, Germany

Printed by Libri Plureos GmbH
in Hamburg, Germany